WITHDRAWN
from the
BT Library
Adastral Park

BT COMMUNICATIONS TECHNOLOGY SERIES 1

# Carrier-scale IP networks:
designing and operating
Internet networks

**Other volumes in this series:**

Volume 2 **Future mobile networks: 3G and beyond** A. Clapton
Volume 3 **Voice over IP** R. Swale

# Carrier-scale IP networks:
designing and operating
Internet networks

Edited by
**Peter Willis**

The Institution of Electrical Engineers

Published by: The Institution of Electrical Engineers, London,
United Kingdom

© 2001: BT EXACT TECHNOLOGIES

This publication is copyright under the Berne Convention and the Universal Copyright Convention. All rights reserved. Apart from any fair dealing for the purposes of research or private study, or criticism or review, as permitted under the Copyright, Designs and Patents Act, 1988, this publication may be reproduced, stored or transmitted, in any forms or by any means, only with the prior permission in writing of the publishers, or in the case of reprographic reproduction in accordance with the terms of licences issued by the Copyright Licensing Agency. Inquiries concerning reproduction outside those terms should be sent to the publishers at the undermentioned address:

The Institution of Electrical Engineers,
Michael Faraday House,
Six Hills Way, Stevenage,
Herts. SG1 2AY, United Kingdom

While the authors and the publishers believe that the information and guidance given in this work are correct, all parties must rely upon their own skill and judgment when making use of them. Neither the authors nor the publishers assume any liability to anyone for any loss or damage caused by any error or omission in the work, whether such error or omission is the result of negligence or any other cause. Any and all such liability is disclaimed.

The moral right of the authors to be identified as authors of this work have been asserted by them in accordance with the Copyright, Designs and Patents Act 1988.

**British Library Cataloguing in Publication Data**

A CIP catalogue record for this book
is available from the British Library

**ISBN 0 85296 982 1**

Typeset in the UK by Mendez Ltd, Ipswich
Printed in the UK by T.J. International, Padstow

# CONTENTS

|   | **Preface** | xi |
|---|---|---|
|   | **Contributors** | xv |
| 1 | **The challenges in building a carrier-scale IP network** | 1 |
|   | *P J Willis* | |
|   | 1.1  Introduction | 1 |
|   | 1.2  Carrier-scale abilities | 2 |
|   | 1.3  What of the future? | 6 |
|   | 1.4  Conclusions | 7 |
| 2 | **An introduction to IP networks** | 9 |
|   | *S Challinor* | |
|   | 2.1  Introduction | 9 |
|   | 2.2  The history of IP | 9 |
|   | 2.3  Fundamentals of IP and overview of packet switching | 11 |
|   | 2.4  Routers | 14 |
|   | 2.5  Routing protocols | 15 |
|   | 2.6  Network technologies | 16 |
|   | 2.7  Domain name system | 18 |
|   | 2.8  IP addressing | 20 |
|   | 2.9  Enhancing IP services | 20 |
|   | 2.10 Internet service provision | 21 |
|   | 2.11 Conclusions | 22 |
| 3 | **Scaling global Internet networks** | 25 |
|   | *A Howcroft* | |
|   | 3.1  Introduction | 25 |
|   | 3.2  Design requirements | 25 |
|   | 3.3  Basic Internet design techniques | 26 |
|   | 3.4  Network Scaling | 36 |
|   | 3.5  Conclusions | 41 |

| | | | |
|---|---|---|---|
| **4** | **The art of peering** | | **43** |
| | *S Bartholomew* | | |
| | 4.1 | What is 'peering' - an introduction | 43 |
| | 4.2 | Tier-1 carriers | 47 |
| | 4.3 | To transit or peer | 47 |
| | 4.4 | Peering dynamics | 49 |
| | 4.5 | Peer with whom and how | 52 |
| | 4.6 | Peering policy and process | 54 |
| | 4.7 | Conclusions | 54 |
| **5** | **How to physically build and maintain PoPs around the world** | | **55** |
| | *T Brain* | | |
| | 5.1 | Introduction | 55 |
| | 5.2 | Location | 55 |
| | 5.3 | The environment | 57 |
| | 5.4 | Power options | 57 |
| | 5.5 | Protective and silent earths | 58 |
| | 5.6 | Earthquake bracing | 58 |
| | 5.7 | Bandwidth ordering | 59 |
| | 5.8 | Equipment ordering | 59 |
| | 5.9 | Equipment delivery | 61 |
| | 5.10 | Equipment installation | 61 |
| | 5.11 | Sparing | 62 |
| | 5.12 | Operational access | 62 |
| | 5.13 | Operational cover | 63 |
| | 5.14 | Conclusions | 64 |
| **6** | **UK core transmission network for the new millennium** | | **65** |
| | *I Hawker, G Hill and I Taylor* | | |
| | 6.1 | Introduction | 65 |
| | 6.2 | Drivers | 66 |
| | 6.3 | Current transport network | 68 |
| | 6.4 | New network developments | 71 |
| | 6.5 | Reducing unit costs | 74 |
| | 6.6 | The optical layer | 76 |
| | 6.7 | Transport for IP services | 78 |
| | 6.8 | Developments in metropolitan networks | 82 |
| | 6.9 | Network management | 83 |
| | 6.10 | Future network management issues | 88 |
| | 6.11 | Conclusions | 88 |

Contents  vii

**7    Delivery of IP over broadband access technologies**                        91
*M Enrico, N Billington, J Kelly and G Young*
    7.1    Introduction                                                          91
    7.2    A brief comparison of broadband access alternatives                   92
    7.3    Near-term evolution of BT's broadband platform                        99
    7.4    The roles of IP and ATM in broadband access systems                  105
    7.5    Delivery of services over BT's broadband platform                    107
    7.6    The evolution of broadband wireline access networks                  114
    7.7    Conclusions                                                          117

**8    Wireless access**                                                          119
*C J Fenton, B Nigeon, B Willis and J Harris*
    8.1    Introduction and scene setting                                       119
    8.2    Wide area public cellular developments                               120
    8.3    Cordless access developments                                         129
    8.4    Fixed wireless access developments                                   135
    8.5    Solutions for wireless access to a carrier-scale IP network          139
    8.6    Conclusions                                                          140

**9    Dial access platform**                                                     143
*J Chuter*
    9.1    IP dial-up service components - introduction                         143
    9.2    Evolution of network provider role                                   144
    9.3    VPN technology                                                       146
    9.4    Platform scaling                                                     147
    9.5    Standards                                                            154
    9.6    Conclusions                                                          155

**10    An overview of satellite access networks**                                157
*M Fitch and A Fidler*
    10.1   Introduction                                                         157
    10.2   Attributes and services                                              159
    10.3   IP access delivery via satellite                                     161
    10.4   System integration                                                   164
    10 5   Conclusions                                                          169

viii   *Contents*

| 11 | **Operations for the IP environment - when the Internet became a serious business** | 171 |
|---|---|---|
| | *J Ozdural* | |
| | 11.1   Introduction | 171 |
| | 11.2   Operations organisation | 172 |
| | 11.3   Functions that need to be performed by operations | 173 |
| | 11.4   The roles | 175 |
| | 11.5   Tools used | 178 |
| | 11.6   Service surround | 185 |
| | 11.7   Conclusions | 187 |
| | Appendix —Investors in people | 187 |
| 12 | **Operational models and processes within BT's and Concert's ISP business** | 189 |
| | *J Meaneaux* | |
| | 12.1   Introduction | 189 |
| | 12.2   What are operational models and processes | 192 |
| | 12.3   ISP organisational development and operations evolution | 192 |
| | 12.4   People development | 194 |
| | 12.5   Conclusions | 195 |
| 13 | **Operational support systems for carrier-scale IP networks** | 197 |
| | *C B Hatch, P C Utton and R E A Leaton* | |
| | 13.1   Introduction | 197 |
| | 13.2   The current approach to provision of IP services | 200 |
| | 13.3   Trends in IP OSS design, development and implementation | 203 |
| | 13.4   Design considerations for an IP OSS | 205 |
| | 13.5   Evolution of networks and services architecture | 208 |
| | 13.6   Conclusions | 214 |
| 14 | **IP address management** | 217 |
| | *P A Roberts and S Challinor* | |
| | 14.1   Introduction | 217 |
| | 14.2   IP addresses | 217 |
| | 14.3   Reserved ranges of IP addresses | 220 |
| | 14.4   IP address co-ordination | 220 |
| | 14.5   IP assignment requests | 224 |
| | 14.6   LIR's address space | 226 |
| | 14.7   Address management tools | 227 |
| | 14.8   Address conservation | 229 |
| | 14.9   Futures | 230 |
| | 14.10  Conclusions | 230 |
| | Appendix —An example of the RIPE-141 form | 231 |

| 15 | **Traffic engineering** | | 235 |
|---|---|---|---|
| | *S Spraggs* | | |
| | 15.1 | Introduction | 235 |
| | 15.2 | Problem definition | 236 |
| | 15.3 | MPLS TE framework | 237 |
| | 15.4 | Traffic engineering QoS | 250 |
| | 15.5 | Traffic engineering configuration | 251 |
| | 15.6 | Implementing traffic engineering | 251 |
| | 15.7 | Conclusions | 259 |
| 16 | **IP virtual private networks** | | 261 |
| | *S Hills, D McLaughlin and N Hanafi* | | |
| | 16.1 | Introduction | 261 |
| | 16.2 | Tag switched VPNs | 261 |
| | 16.3 | IPSec VPNs | 265 |
| | 16.4 | Which technology? | 274 |
| | 16.5 | Conclusions | 278 |
| | **Acronyms** | | 281 |
| | **Index** | | 289 |

# PREFACE

The aim of this book is to give the reader an understanding of all the aspects of designing, building and operating a large global IP network, including an overview of many of the transport and access components of an IP network. The contributions are written by world-class technical experts and seasoned IP network managers. This should give any reader studying, or planning to build and operate, a large IP network a flying start.

This book divides the subject of Carrier-scale IP Networks into the following five areas:

- overview;
- designing and building IP networks;
- transmission and access networks;
- operations;
- development of future networks.

Network Management is covered in individual chapters where appropriate.

The first part provides an overview of carrier-scale networks and of IP networking in general. Chapter 1 explains some of the attributes that contribute to successful carrier-scale IP networks. Chapter 2 gives an introduction to basic IP and Internet terminology. The author of this chapter designed BT's IP backbone, the largest IP backbone in the UK, but more importantly for the reader he has the skill to describe the fundamentals of IP networks in a way everybody can understand. Those already familiar with IP and the Internet may wish to skip Chapter 2.

The second part describes the scaling issues in IP networks, how they peer with other networks and some of the practicalities of building and maintaining them. Chapter 3 explains how to design global Internet networks so that they scale, taking into account the fundamental limits of routers and routing protocols. The author has been designing data networks since 1985 and designed BT's first European Internet backbone network in 1995; he is currently designing extensions to Concert's global Internet backbone. This chapter is therefore an authoritative source on the design methodology used during the planning and growth of BT's and Concert's IP backbones. Chapter 4 shows how a key to offering good Internet service is providing good connectivity to the rest of the Internet. The art of peering is connecting to the other Internet providers at best cost and maximum performance. The author of this chapter, the current peering co-ordinator for Concert, decided to call this chapter 'The Art of Peering' because Internet peering is much more than making technical decisions, although the technical decisions are significant, it is also about commercial policies and relationships. Chapter 5 shows how, in an ever-growing network, finding the right kind of accommodation and installing the network kit at a hectic rate provides a real physical challenge. The author manages Concert's IP planning team and therefore lives every day with the challenges of building and upgrading PoPs on a global basis.

The third part deals with the transmission and access networks that provide the connections between the points of presence of an Internet provider and their customers. Chapter 6 covers wired access where the network is accessed via any type of wired or cabled mechanism. The chapter shows that one of the basic ingredients of an Internet service is bandwidth and then goes on to discuss how to use fibre, SDH and the new digital transmission techniques to give best-cost bandwidth. Chapter 7 provides an overview of the broadband access technologies that can be used to deliver high-speed IP services to the residential and SME markets, while Chapter 8 covers similar ground where the network is accessed via a wireless, cellular or radio mechanism. Chapter 9 details the architecture and design of BT's huge dial access platform — this is where the customer dials into the network via PSTN or ISDN. Chapter 10 gives details of the various ways satellites can be used to access IP networks where the customer uses a satellite connection for network access.

The fourth part — operations — is less overtly technical but its subjects are still key to the success of carrier-scale networks. Chapter 11 describes how a team of multi-skilled people has to be brought together and managed to efficiently operate a large IP network. The author of this chapter worked as an IP engineering specialist with BT's Internet Operations Centre for several years — this experience endowed him with a pragmatism that is reflected in the title and contents of this chapter. If you want to know what an Internet Operations Centre does or want to know how to develop or organise one then read this chapter. Chapter 12 describes how IP network production models have evolved and how they influence the development of systems and people. Key to operating a carrier-scale network are the processes and procedures used to operate the network and provide services — the larger the network, the more customers it has, and the more dependent on good processes the service becomes. Chapter 13 describes the operational support services that provide the tools and automation to implement the processes required to operate the network and deliver service to customers. Grouped in this section is Chapter 14 on IP address management. IP addresses are key to the operation of an IP network. IP addresses have to be carefully managed following rules laid down by the IP registries, the providers of all IP addresses. Readers wishing to design IP networks or who need to know how to obtain and manage IP addresses should read the complete chapter, while readers who require just an overview of IP addresses may want to read only sections 14.1 and 14.2.

The last part covers some key aspects of the development of future networks. Chapter 15 describes how traffic flows are manipulated across the network in order to reduce congestion and obtain more value for money from under-utilised lines. It discusses traffic engineering and how the new multi-protocol label switching technology can be used to improve the cost performance of an IP network. This is a technical chapter and aimed at those readers who wish to understand how to use MPLS to traffic engineer networks. The author, being a Cisco engineer, has focused particularly on that company's views of traffic engineering. Chapter 16 shows how Internet technology is now being used in private networks. The Internet is a public network with performance and security implications a private company may want to avoid. Virtual private network technology allows secure networks with guaranteed levels of performance to be delivered on a shared infrastructure.

Finally, I would like to thank all the authors for what is collectively a most comprehensive overview of carrier-scale IP network design, provision and operations.

Peter Willis
IP Networks Architect, BTexact Technologies
peter.j.willis@bt.com

# CONTRIBUTORS

S Bartholomew, IP Programmes, Concert

N Billington, Access Solution Design, BTexact Technologies

T Brain, IP Network Planning, Concert

S Challinor, IP and Data Service Design, BT Ignite

J Chuter, IP Solutions, BTexact Technologies

M Enrico, formerly Access Solution Design, BTexact Technologies

A Fidler, Satellite Systems, BTexact Technologies

M Fitch, Satellite Network Development, BTexact Technologies

C J Fenton, 3G Service Development, BT Wireless

N Hanafi, Intranet Services, Concert

J Harris, VoIP Network Solutions, BTexact Technologies

C B Hatch, Systems Evolution, BTexact Technologies

I Hawker, formerly Systems Engineering, BTexact Technologies

G Hill, formerly Systems Engineering, BTexact Technologies

S Hills, Interworking, BTexact Technologies

A Howcroft, Technology Business Development, Concert

J Kelly, formerly Access Solution Design, BTexact Technologies

R E A Leaton, OSS Design, BTexact Technologies

D McGlaughlin, Security Design, BTexact Technologies

J Meaneaux, formerly Firewall Products, Concert

B Nigeon, Wireless Home Networking, BT Retail

J Ozdural, Business Technical Architecture, Syncordia Solutions

P A Roberts, Technical Support, Syncordia Solutions

S Spraggs, IP Networks Design, Cisco Systems

I Taylor, Systems Engineering, BTexact Technologies

P C Utton, IP Network Management, BTexact Technologies

B Willis, IP and Data Networks Technology, BTexact Technologies

P J Willis, IP Networks Architect, BTexact Technologies

G Young, formerly Access Solutions Design, BTexact Technologies

# 1

# THE CHALLENGES IN BUILDING A CARRIER-SCALE IP NETWORK

## P J Willis

## 1.1 Introduction

This chapter describes what is meant by a carrier-scale IP network. It details the challenges in designing, building and operating a carrier-scale IP network, and lists the abilities required in a network to make it carrier-scale, e.g. scalable, manageable, performant and secure. Also described are the challenges for the future when developing an esoteric protocol or service, such as multicast or VoIP, into a carrier-scale protocol or service. None of the challenges highlighted here should be underestimated in their capability to confound, and they are best met by applying the solid design and operations principles covered in the following chapters.

Firstly, the term 'carrier-scale IP network' has to be defined before its properties and challenges can be described. A 'carrier-scale IP network' is an IP network with the following properties:

- it provides services for millions of end users;
- it provides high-speed, greater than 100 Mbit/s, transit services to other Internet service providers (ISPs) — in a traditional retail/wholesale model this would be considered a wholesale service;
- it is reliable, scalable and manageable.

Today's best Internet providers demonstrate these features and tomorrow's IP virtual private networks (VPNs) will have these features.

Many challenges face the designers, engineers, builders and operators of large global IP networks. The challenges range from the physical challenges of finding enough accommodation with enough power and air-conditioning, to the more abstract challenge of configuring the network in such a way that it performs well and operates reliably. Configuring a network can be compared to writing a computer program — one wrong line of configuration code can bring down the whole network. Unlike conventional computer programming the network designer has the additional challenge of making the system stable; therefore, although a network design may work logically, in the real world, with speed-of-light constraints and finite processing capabilities, the design might not be stable.

The challenge of carrier-scale IP networks is more than designing and building a network that functions — it is designing and building a network with abilities that make it possible to grow, manage and operate it. These abilities can be described as carrier-scale abilities.

## 1.2 Carrier-Scale Abilities

Carrier-scale abilities are the abilities of a network to be:
- scalable — this is the ability of the network to grow, i.e. it must be possible to increase the number of points of presence (PoPs) and the number of routers, grow the size of the circuits and to uniquely address all its components and customers;
- manageable — it must be possible to manage the network;
- performant — the network must perform adequately and it must be possible to measure the performance of the network and engineer its performance to meet service level agreements;
- secure — the components of the network need to be secure from attack.

The following sections describe the above abilities in more detail.

### 1.2.1 Scaling Networks

IP networks cannot simply be made larger by adding more routers or made to go more quickly by adding more or faster circuits. There are a number of constraints a network designer must consider when designing large IP networks:
- physical space, number of connectors per router;
- power consumption and dissipation;
- processing power and memory size of individual routers;
- switching speed of individual routers;
- limits of the routing protocols used to control the network.

Chapter 3 describes how to design a large IP network and the key criteria to scaling an IP network.

The essential design rule for scaling IP networks is to make them hierarchical. In a hierarchical network customers connect to access routers. Access routers connect to core routers. The PoPs are arranged hierarchically so that small PoPs connect to medium size PoPs that connect to larger PoPs, where the size of a PoP is a relative measure of the aggregate amount of bandwidth of the circuits connected to that PoP. The size and shape of the router hierarchy is determined by consideration of the size of the network and its traffic patterns, both of which can be difficult to predict for a new network. Network designers therefore have to build an ability to flex the shape and size of the hierarchy into the design.

### 1.2.2 Managing Networks

There are two main management categories:
- service management — this is the management of services sold to customers;
- network management — this is the management of the functions of the network.

#### 1.2.2.1 Service Management

Service management deals with the general services sold to customers, e.g. delivery of service, taking orders, measuring and reporting on service performance. Service management is driven by processes and supported by operational support systems (OSS).

An important consideration in defining processes is how rigorously do they need to be defined. For complex and highly customised advanced services, processes need to be lightweight. For mass-production mature services, the processes need to be rigorously defined, accurate and proven. It is important to recognise that products and services mature, and as the products mature the processes should be refined. For example, when BT first launched Internet products they were seen as complex and were delivered using lightweight processes. Over time as the understanding of the products improved the processes and support systems were refined to eventually enable BT to deliver Internet products to thousands of customers very efficiently with rigorous processes. Understanding the life cycle of IP services, the need to run with both lightweight processes for new products and rigorous processes for mature products, at the same time and in the same operation centres, is key to understanding IP service management.

Operation centres have to be organised to deal with customers' requests and queries as quickly as possible and to make effective use of highly skilled IP people. To achieve this the operation centre is organised in tiers or lines; the first line are the people who are initially contacted by the customer — this is by telephone or e-mail. The first line will be equipped to deal with a large number of queries and trained to handle the simpler day-to-day requests. Any request that cannot be handled by the first line will be passed on to a more technically skilled second line. Several lines or tiers may exist. The art is to ensure the skills of the higher tiers are focused on designing and improving the IP services while at the same time supporting the more technically demanding customer requests.

#### 1.2.2.2 Network Management

Network management is divided into the following five areas.

- Configuration management

  This is the ability to configure the network, monitor the configuration, change the configuration and roll back to previous configurations in the event of a failure.

- Fault management

  This is the ability to monitor the network for faults and fix any faults discovered or reported by customers.

- Performance

  This is the ability to monitor, model and tune the performance of the network. A key element of performance management is traffic engineering, the ability to manage where traffic flows — see Chapter 15.

- Security

  This is the ability to secure the devices in the network and monitor them for attacks.

- Accounting

  This is the ability to gather information about the usage of the network by customers. If the customers are not billed for usage then this area may not be important.

When designing a network or adding new components or new protocols to an existing network, the above areas of network management must be addressed. If the component of design is deficient in any of these areas, it is not manageable.

These important areas of service and network management are covered in more detail in Chapters 11, 12 and 13.

### 1.2.3  Dealing with the Rate of Change

A perceived problem in designing and operating IP networks is dealing with the rate of change of technology. Many people have heard of 'Internet years', where one real year is equivalent to 5—7 Internet years. It appears that every week we hear stories of new mergers, new Internet applications and faster, more functional network boxes (e.g. terabit routers). While there is a technology race to improve the speed of routers and add more functionality to them, the basic principles of the Internet have remained unchanged over many years — see Chapter 2 for a review of the history and basic principles of the Internet. The most significant change in recent times was the introduction of classless inter-domain routing (CIDR) and BGP version 4 in 1994. Changes currently affecting the Internet are quality of service (QoS) mechanisms and multicast. The biggest challenge facing IP networks today is coping with the rate of growth. Estimates for the rate of growth of individual Internet service providers range from doubling every year to increasing by a factor of ten every year. BT's UK Internet infrastructure has consistently experienced a 400% per year growth in terms of bandwidth and customers since its launch in 1994. To deal with such large growth rates the scaling and management features of BT's networks are key.

### 1.2.4   Network Security

The components of a network, for example a router, need to be secure. Aspects of network security are:

- making the components of the network secure from attack — this is done by ensuring that only identified management systems have access to the management interfaces of the network components; there may be capabilities which equipment vendors put in certain items of network equipment to make them easier to manage in a secure enterprise environment, but which are not suitable for an insecure public Internet environment — these capabilities must therefore be turned off;

- detecting an attack — heuristic methods can be used to determine the differences between the legitimate attempts by network management people to access equipment, and those by attackers; similarly heuristic methods can be used to determine if equipment is being attacked, by scrutiny of appropriate logs of the equipment's activities;

- knowing your own vulnerabilities — network equipment can be checked by security-checking software to test for vulnerabilities; network operators should also ensure that their equipment vendors notify them of bugs that might affect security;

- controlling management access rights carefully — as a network might be attacked by an insider, it is important not to grant access rights to everybody in network operations, but only sufficient rights to each individual to enable that person to perform their identified role; it is also important that, as individuals change jobs or leave network operations, their management rights are changed or revoked in a timely manner;

- shutting-off attackers — have plans to deal with attackers, e.g. by ensuring that an attacker can be cut-off to prevent them doing any more damage; this may require co-ordination with other networks or agencies;
- undoing an attacker's damage — configuration management systems are required that can restore the network configuration in the event that the attacker has managed to change the configuration.

## 1.3  What of the Future?

This chapter (and the rest of this book) has not ventured into the realms of IP technologies that are, at the time of writing, esoteric, such as multicast, VoIP, protocols and systems to provide QoS, and the wide range of protocols to support mobility, e.g. mobile IP. A further book could be produced just dedicated to these esoteric protocols and systems, describing how they could be used in future carrier-scale IP networks. This section will just cover the challenges in developing an esoteric protocol into a carrier-scale protocol:

- designers understanding the new protocol and its capabilities;
- product lines understanding the new services and options available to customers;
- understanding the scaling abilities of new protocols, e.g. the maximum size of the network that the new protocol will work across, and whether it will be stable in large networks or quickly changing networks — it may not be possible to rework the protocol to make it stable but the stability may be improved by tuning configurable settings such as timers;
- understanding the impact of the protocols on the resources of the network, e.g. how much router CPU it consumes;
- understanding the performance of the new protocol, e.g. how many sessions per second it can handle;
- understanding the management features of the new protocol — what needs to be managed and how it will be managed;
- understanding the vendors' capabilities to support this protocol, e.g. if it is only currently supported in development pre-release code, when the protocol software will be fully supported, how reliable, how well debugged the vendors' software is that implements this protocol;
- defining the processes for operating and maintaining this protocol in the network, due consideration being given to the development of the processes over the protocol's lifetime — while in the early days the new protocol may be supported by experts, through a process of training programmes, new tools and operational experience, the processes will be refined and the maintenance of the new protocol passed to mainstream operations;

- understanding and defining the roll-out of the new protocol — whether it will be deployed everywhere on day one of its launch in the network or only be initially deployed in a small number of locations;
- training operations people to support the new protocol;
- developing new OSS and new NMS capabilities to support the new protocol;
- defining the new products that could be sold to customers;
- defining the processes to sell and deliver the new products to customers, bearing in mind that processes are evolutionary;
- assessing to what level the customer will take up the new product, and how much human and system resource will be required to meet the expected customer demand.

It should be noted that none of the above challenges involves modifying the protocol directly — one of the more difficult challenges is having the ability and skill to influence protocol developments, particularly the IETF work, to ensure that only protocols possessing carrier-scale abilities are defined and implemented. It is interesting to note that the only protocol designed to enable IP to become more carrier scale is the set of protocols known as IPv6. From the outset IPv6 was intended to allow the Internet to become much larger and more manageable through the use of more automation of configuration, higher performance and greater security. The introduction of IPv6 into today's carrier-scale IP networks may never happen unless all the challenges listed above can be overcome.

New services and protocols which may be introduced over the next few years include the introduction of a new transport protocol called MPLS, more QoS features, more network-based features and more address space with IPv6. While these changes are happening, the basic principles and the good engineering practices required to design, build and operate IP networks will not change.

## 1.4  Conclusions

A quote, which comes from 'Alice Through The Looking Glass', reflects on the dilemma the designers, builders and operators of carrier-scale IP networks have:

> 'A slow sort of country!' said the Queen. 'Now, here, you see, it takes all the running you can do, to keep in the same place. If you want to get somewhere else, you must run at least twice as fast as that!'

Despite the need to run and change fast, the solid principles of design and operation overviewed in this chapter still apply. Only if the correct scaling and management features are built into the network and its operations will it successfully cope with the rate of growth and change.

# 2

# AN INTRODUCTION TO IP NETWORKS

### S Challinor

## 2.1 Introduction

This chapter is an easy-to-read introduction to the technology and history of IP networks. It provides a condensed summary of IP terms and concepts essential to the understanding of the rest of this book. Those with a basic understanding of IP may be tempted to skip this chapter but some may find that it covers gaps in their knowledge.

The chapter introduces the reader to IP networking, covering the fundamental principles of packet switching, through to the devices, protocols, technologies and applications used to create an IP network.

## 2.2 The History of IP

IP, the Internet protocol, stems from work started in the US Department of Defense in the late 1960s to create a network resilient enough to withstand an enemy attack.

This network, which initially linked centres of defence and academic institutions, eventually became the Internet. Figure 2.1 illustrates the dramatic growth of the Internet, and some of the milestones along the way.

Some milestones are listed below.

**1969** — US Department of Defense Advanced Research Projects Agency network connecting four universities in the United States: Stanford Research Institute, UCLA, UC Santa Barbara, and the University of Utah. The network is used for applications such as e-mail and file transfer.

**1973** — First international connections to Arpanet University College, London, and the Royal Radar Establishment in Norway.

10  *The History of IP*

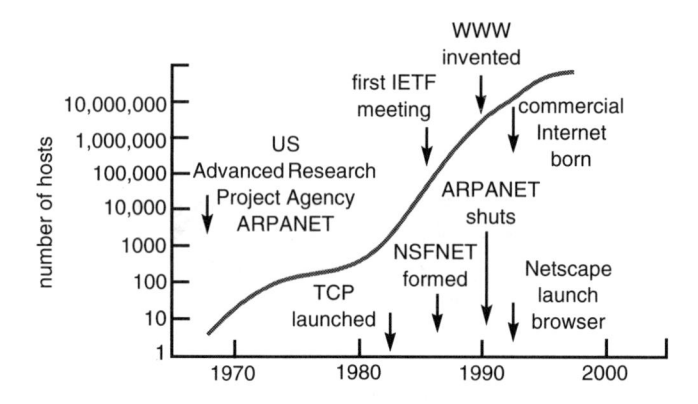

**Fig 2.1**  The growth of the Internet.

**1979** — First appearance of 'Usenet', an on-line discussion group via e-mail.

**1982** — TCP/IP created, the underlying protocol of all Internet computers. The term 'Internet' is used for the first time.

**1986** — The first IETF (Internet Engineering Task Force) meeting. The IETF is the protocol engineering, development, and standardisation arm of the Internet Architecture Board (IAB). Its roles include identifying, and proposing solutions to, pressing operational and technical problems in the Internet, specifying the development or usage of protocols and the near-term architecture to solve technical problems for the Internet, making recommendations to the IAB regarding standardisation of protocols and protocol usage in the Internet, facilitating technology transfer from the Internet Research Task Force (IRTF) to the wider Internet community, and providing a forum for the exchange of information within the Internet community between vendors, users, researchers, agency contractors, and network managers. IETF standards are made available as 'Requests for Comment' or RFCs. RFCs in draft are referred to as 'Internet drafts'

**1986** — US National Science Foundation links five super computer sites to form the NSFnet. This becomes the backbone to the Internet, but does not carry commercial traffic.

**1990** — The Arpanet shuts, leaving the Internet it created as a thriving network of networks with 300 000 hosts and growing rapidly.

**1991** — First appearance of the World Wide Web.

**1993** — Restriction of commercial use of US backbone (NSFnet) is removed.

**1993** — First Web browser, Mosaic, soon to be followed by Netscape — a company set up by one of the original designers of the World Wide Web, Marc Andreesen.

**1995** — NSFnet reverts to being a research project, leaving the Internet backbone in commercial hands.

**1996** — Approximately 40 million people connected to the Internet.

## 2.3 Fundamentals of IP and Overview of Packet Switching

The traditional telecommunications voice network provides a dedicated circuit between two telephones for the duration of a call. The route of the circuit through the network is fixed at the beginning of the call, and no other calls may use the bandwidth reserved by a call in progress, even if nobody is actually speaking. Datacommunications networks operated on similar principles — predetermined circuits through the network for data traffic, with all or part of the bandwidth on the circuit guaranteed.

IP turns this model around completely. It should not be thought of as the next contender in a history of datacommunications technologies, but instead as a fundamental change in the way that networks are viewed. Three key ideas underpin IP networks:

- the 'bomb-proof network' — IP's origins in the world of defence mean that IP has been designed to connect two end-points, even though some of the points in between may be lost at any time;

- the end-to-end principle — in its simplest form, an IP network only transmits IP packets from one point to another (all functions that may be performed outside the network should be performed outside the network);

- IP over everything — internetworking is best achieved by layering a unique internetworking protocol on top of the various different network technologies — thus IP does not replace other networking technologies, it links them together into a single network.

IP uses 'connectionless routing' or 'packet switching' to carry traffic between two points. Traffic is broken down into a number of discrete blocks called 'datagrams', and each datagram is individually addressed with the destination before being passed into the network. The datagrams are passed from node to node through the network, with each node examining the destination address to decide where next to send the packet, this is sometimes called 'hop-by-hop' routing. No bandwidth is reserved for the packets, and there is no guarantee that two packets going from the same source to the same destination will follow the same route. It can be seen that packet switching is analogous to the postal network, where parcels and letters enter the system, each with the address written on the front, and are then sorted at intermediate post offices until they reach their destination. A single message divided up and posted as two letters going to the same address may take different routes through the network, and may even arrive in a different order to the order in which they were posted. At their destination they may be opened and the content re-assembled to form the full message. Figures 2.2 and 2.3 illustrate this comparison.

The fundamental ideas of IP and the packet-switching model offer several key benefits over a traditional circuit-switching network.

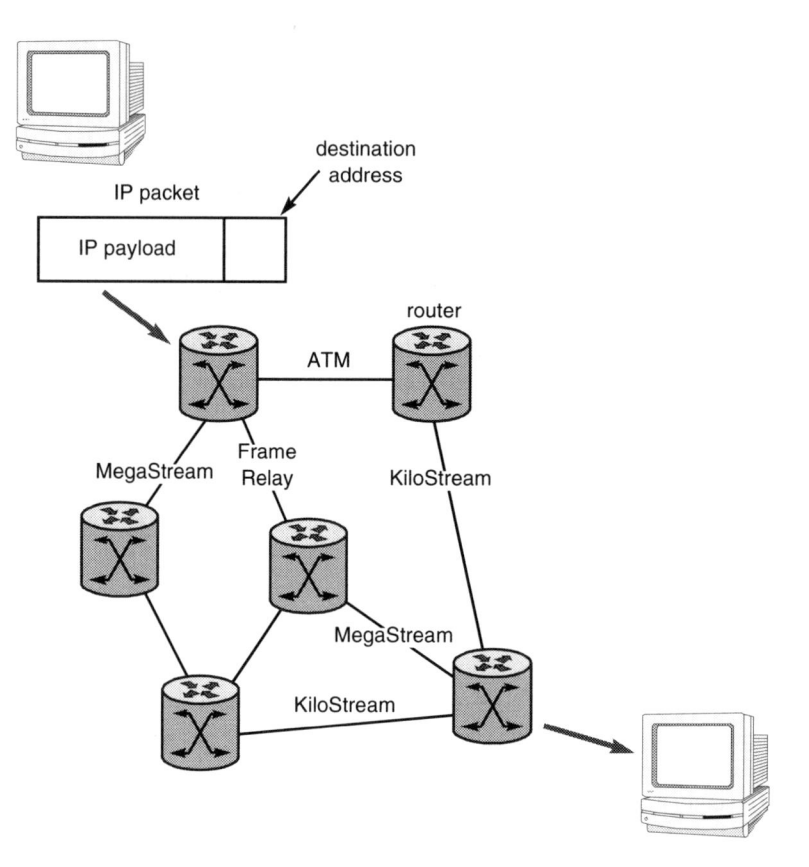

12  *Fundamentals of IP and Overview of Packet Switching*

**Fig 2.2**  A simple IP network.

- No need for reserved bandwidth

  Without a need to reserve bandwidth for sessions that do not require it all the time, less network resource is required to support the same volume of traffic, thus reducing cost. It should be noted, however, that, as packet switching does not offer a comparable service to a circuit-switched network without a considerable amount of extra work, the cost savings are therefore only realised when much of the traffic does not actually require a dedicated amount of bandwidth. It is beginning to be possible to reserve bandwidth on IP networks for traffic types that do require it.

- Resilience

  Since IP datagrams make fresh routing decisions at every hop through the network, IP networks are very good at dealing with link failures — the IP packet simply takes the next best alternative route.

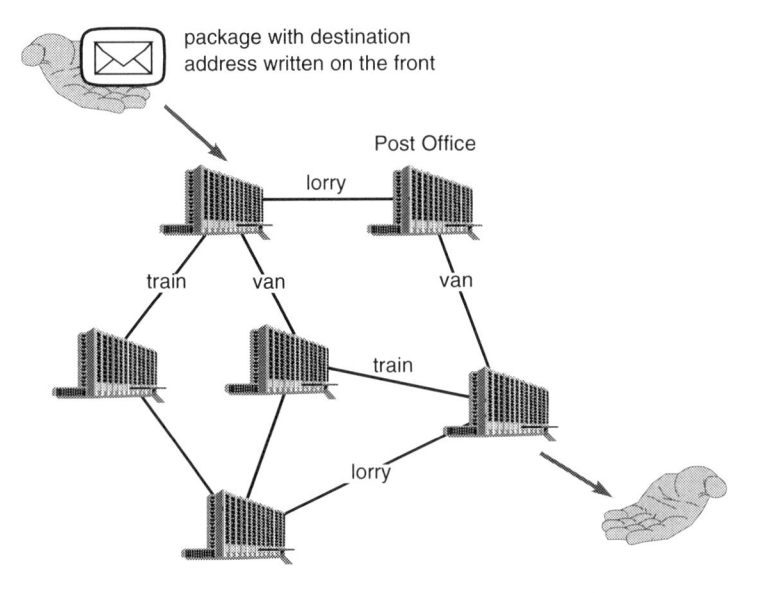

**Fig 2.3** IP network analogy — a postal network.

- Network technology independence

    IP packets provide a method of carrying data over completely different technologies. The IP packet may be carried across several different network types on its journey from source to destination in the same way as a postal letter may use cars, lorries and trains as it crosses the country hopping between postal sorting offices (see Figs 2.2 and 2.3).

    Since there is no guarantee that an IP packet will be delivered by the network, it is the responsibility of the sending and receiving hosts to determine that an IP packet has been delivered correctly. This is performed by the transmission control protocol (TCP). TCP handles the speed at which IP packets are fed on to the network, sends acknowledgements for packets received, checks for acknowledgements of packets sent and ensures that packets received are reassembled in the correct order. If the post office analogy is continued, TCP may be considered to be the secretaries of the sending and receiving parties checking with each other that the packages sent have all been received and are intact.

    IP sessions that do not require this level of checking, e.g. a voice call, may use the light-weight unacknowledged datagram protocol (UDP) — UDP may be considered analogous to a bad secretary, who sends packages, but never checks to see if they have arrived, and does not care!

    Packet-switched networks are often referred to in terms of three layers. The term 'layer 1' is used to describe the physical wires/fibres involved in building the network. 'Layer 2' refers to virtual connections between end-devices and routers —

the low-level protocols that control how data is actually encoded on the wire, and how a single physical connection into a router may be shared between many customers. 'Layer 3' determines how packets are routed between the sender and the receiver — this is the layer in which the Internet protocol operates.

## 2.4 Routers

Routers are the main building blocks of an IP network. They perform the following key functions.

- Routing traffic

    Traffic arriving on one link into a router is analysed to determine its destination. The router checks its routing table to determine the next hop for that destination, and then transmits the packet over the link to that next hop.

- Building routing tables

    Routers may be manually configured with information on how to route traffic to given destinations; however, most networks run a 'routing protocol' to communicate information about the network topology to each other, thus allowing them to determine the best next hop for a destination.

- Connecting different network technologies

    A router encapsulates the IP packet in the format required of the underlying network, e.g. a frame relay network may be connected to an Ethernet network by a router which de-encapsulates a packet from its frame relay frame, then re-encapsulates it into an Ethernet frame. In this respect, a router may be thought of as a post office, taking packets from a lorry, sorting them, and loading them on to a train.

Routers exist in many shapes, sizes and configurations, each suitable for a particular role in the network. A brief summary of today's router types and the role they may perform within a large IP network is presented below.

- Small routers

    Typically used as customer premises equipment (CPE), they have a couple of local area network (LAN) and a couple of wide area network (WAN) interfaces and typically support links of around 2 Mbit/s (see section 6 for definitions of LAN and WAN).

- Average-sized routers

    Typically used as access routers in the service provider's network, they terminate access lines from around 200 customers and support WAN links of around 34 Mbit/s.

- Large routers

    Typically found in the core, they support mainly the high-speed networking technologies and carry aggregated traffic from large numbers of customers. Link speeds of up to 155 Mbit/s would be supported.

- Very large routers

    New routers are constantly being developed, each offering huge capacity advantages over the previous generations. Gigabit routers are now widely available that support 2.4 Gbit/s links and have a total switching capacity of up to 60 Gbit/s. Such routers are now appearing in the core of large IP networks. Routers capable of switching terabits of information are expected to become available over the next couple of years.

    The physical location that houses a router, or a number of routers, is usually referred to as a 'PoP' — a point of presence.

## 2.5 Routing Protocols

In order to route IP packets towards their destination, routers must maintain a routing table to map destination networks to the best next hop. Routes may be manually defined into this routing table using a 'static' route. Although conceptually straightforward, static routing has several disadvantages. It is time-consuming to manually configure, there is a great risk of configuration errors and it offers minimal resilience to failure or topology changes.

If the routers themselves are to determine the route to a destination, they must run a 'routing protocol' to build the routing table, this is called 'dynamic routing'. The routing protocol defines the way that routers communicate to their neighbours to build a picture of the network and the status of the links. Under failure conditions, or when a link is added or removed from the network, it is the routing protocol that allows the routers to re-build their routing tables. Once set up, dynamic routing requires little manual configuration and is able to cope with changes and failures; however, it does require additional processing to be performed by the routers and consumes a small amount of network bandwidth. The routing protocols used in dynamic routing divide roughly into the following two types:

- Interior routing protocols

    These build a picture of the interior of a network, and an exterior routing protocol is used to exchange routing information with other networks — typically networks maintained by other organisations. The interior routing protocols are optimised for fast convergence (i.e. quickly react to changes). Examples include RIP and OSPF.

- Exterior routing protocols

    Optimised for the bulk exchange of routes, they set the policy about what traffic should be routed where. BGP4 is the exterior routing protocol used on the Internet. All the devices under the routing control of one organisation are said to be within the same BGP 'autonomous system' (AS), and the BGP4 protocol is used to convey routing information between these ASs. An ISP will define a 'routing policy', which is a list of rules about how traffic to and from other ASs should be dealt with. It is the routing policy that allows the ISP to alter traffic routing to fulfil its commercial model as well as its technical requirements.

## 2.6 Network Technologies

IP may be carried over many different network technologies. Allowing IP to be carried on just about any existing installed networking infrastructure has the key benefit that network technologies may be selected to suit the requirements of each part of the IP network. For instance, an access link may require a low-cost, low-bandwidth, medium-resilience and long-range connection, whereas a core link may require a high-bandwidth, high-resilience, long-range link with few cost restrictions. Below is a list of some of the key networking terms and technologies with a brief summary of their attributes.

- Leased lines

    These are point-to-point connections — BT sells them as 'KiloStream' and 'MegaStream', with speeds ranging from 64 kbit/s to 155 Mbit/s and beyond. Such links can connect any two points in the world (where service is available) and provide dedicated bandwidth between them, usually at a range of reliability levels. Leased lines are one of the more expensive ways of obtaining connectivity within an IP network, and are usually used only when the full bandwidth is required for most of the time.

- Synchronous digital hierarchy (SDH)

    This is the underlying technology used to provide many of the higher speed leased lines. SDH may be used to support links at a variety of speeds as follows:

    — STM-1 runs at approximately 155 Mbit/s (also called an OC3);

    — STM-4 runs at approximately 622 Mbit/s (also called an OC12);

    — STM-16 runs at approximately 2.4 Gbit/s (also called an OC48);

    — STM-64 runs at approximately 10 Gbit/s (also called an OC192).

    It should be noted that, even though SDH is defined up to 10 Gbit/s, this does not mean that such speeds may be purchased from the service provider for a single leased line (see Chapter 6 for more detail).

- Dense wave division multiplexing (DWDM)

  This is a technology where multiple wavelengths of light are transmitted across a single fibre optic cable instead of just one wavelength, thus greatly increasing the capacity of a fibre. DWDM is typically used to build the infrastructure that provides leased lines to networks. However, large IP core routers are increasingly using DWDM directly as a way of obtaining a large amount of bandwidth.

- Frame relay

  A frame relay network is a single network built using leased lines that may be shared by many completely separate networks. Each network buys a number of 'virtual circuits' through the frame relay network, each circuit usually having a guaranteed minimum bandwidth, and a maximum bandwidth. It would not be possible for all networks to use their maximum bandwidth at the same time, but since this is statistically unlikely to occur, frame relay allows networks to be built that can burst for short periods when required, at a far lower cost. An IP network would typically use frame relay in the access layer to provide many long- or short-range customer links into the IP core. Frame relay is most commonly used for link speeds of up to 2 Mbit/s.

- Asynchronous transfer mode (ATM)

  Superficially, an ATM network may be thought of in similar terms to a frame relay network. Like frame relay, ATM allows many lower cost virtual connections to share a common infrastructure made up of leased lines. ATM adds to the functionality of frame relay by offering a greater range of virtual circuit types (e.g. guaranteed bandwidth, low jitter circuits for voice traffic) at greater bandwidths. The greater bandwidth range means that ATM may be found in the core of average-sized IP networks, as well as at the access layer.

- Ethernet

  This is a network technology used almost exclusively in the local area (thus Ethernet is a LAN technology). It offers a range of bandwidths, from 10 Mbit/s to 1000 Mbit/s, over many different physical implementation methods, ranging from direct coaxial cable through to fibre optics fanning out from a switching Ethernet 'hub' device. It is a cost-effective and flexible way of providing a large amount of bandwidth between devices within a building. The key attribute of Ethernet is its broadcast nature; the technologies so far described are point-to-point technologies, i.e. they provide a link, either real or virtual, between two points. Ethernet, however, may be viewed as a shared 'block' of bandwidth to which a number of devices connect to pass information between themselves, either between just two machines, or from one machine to all the others on the shared medium. Ethernet is normally found linking office computers to file servers and routers. An Internet PoP may use Ethernet within the PoP to connect some or all of the routers together to allow traffic to cross the PoP on route to its destination.

18  *Domain Name System (DNS)*

- Fibre distributed digital interface (FDDI)

  This is a LAN technology where a number of devices are linked together using fibre optics in a ring topology. The ring runs at 100 Mbit/s, and is resilient to a single break anywhere in the fibre.

- Token ring

  A LAN technology that is superficially similar to a low speed Ethernet, token ring runs at 4 Mbit/s or 16 Mbit/s, and differs from Ethernet in the way that it controls a machine's access on to the shared medium.

- Switched multimegabit data service (SMDS)

  This is a long-distance service where multiple devices connect to a single virtual network — in this respect SMDS appears like a large, long-distance Ethernet. SMDS runs at speeds of up to 10 Mbit/s.

- X.25

  Very much a legacy network technology in the IP world, X.25 offers low-speed, high-reliability links.

- Satellite

  Satellite is used for very long distance connections, typically when a land line alternative is not readily available (see Chapter 10 for more information about using satellites in IP networks).

- Digital subscriber loop (DSL)

  This is a technology that allows several Mbit/s of traffic to be carried on the traditional copper local loop (see Chapter 7 for more information about DSL).

- Wireless

  There are a number of technologies that allow data to be carried without the need for a cable. This may be used by a network operator to provide a link in a network across a piece of land where a physical cable is not possible or prohibitively expensive, or it may be used to reach a mobile terminal with a small or large geographic area (see Chapter 4 for use of wireless systems in IP networks).

## 2.7  Domain Name System (DNS)

The Internet as we know it would be completely unusable without the domain name system to convert between IP addresses and server names. When an IP transaction is initiated using a server's name, looking for a Web page for example, the local machine needs to convert the name to an IP address. If it does not hold a local record of the mapping, it must consult a DNS server to find the correct IP address; only then can the transaction continue. DNS may be thought of as a world-wide distributed database made up of thousands of 'DNS servers', each holding name-to-address mappings for the domain they control, and referring to other DNS servers for domains they do not control. The DNS servers are arranged in a hierarchy, a simplified view of which is presented in Fig 2.4.

A worked example of a PC using DNS to look up a Web site is presented below:

- a user on a PC somewhere in the Internet requests the Web page 'www.mynet.bt.co.uk';

- the PC checks to see if it has a local mapping, either manually entered, or cached from a previous look-up for this name;

- the PC requests a DNS server to perform the DNS look-up on its behalf,;

- the DNS server checks its local records, and its local cache of recent look-ups to see if it can answer the query;

- since the DNS server does not have any record of the required name, it queries the root DNS at the top of the hierarchy — the root server supplies the DNS server with the address of a DNS server that knows about the 'UK' namespace;

- the DNS server queries the server authoritative for the 'uk' namespace;

- the server authoritative for 'uk' cannot answer the request directly, but, since it knows about all the DNS servers within the 'uk' namespace, it refers the local DNS server down the hierarchy to the server authoritative for the '.co.uk' namespace;

- the DNS server queries the server authoritative for the '.co.uk' namespace;

- the server authoritative for '.co.uk' cannot answer the request directly, but, since it knows about all the DNS servers within the '.co.uk' namespace, it refers the local DNS server down the hierarchy to the server authoritative for the '.bt.co.uk' namespace;

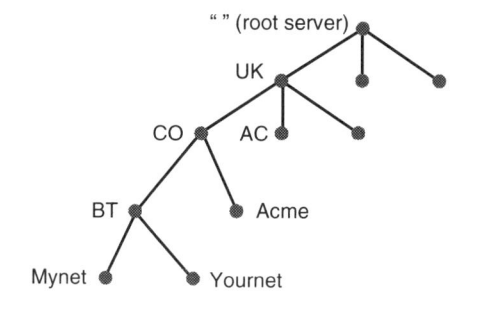

**Fig 2.4** DNS server hierarchy.

- the DNS server queries the server authoritative for the '.bt.co.uk' namespace;
- the server authoritative for the '.bt.co.uk' namespace refers the local server down the hierarchy again to the server which holds information for the 'mynet.bt.co.uk' namespace;
- the DNS server queries the server authoritative for 'mynet.bt.co.uk', and receives a reply containing the correct IP address for the machine called 'www';
- the DNS server returns the reply to the PC, and caches the information for a length of time defined by information contained in the reply (it is possible to specify how long DNS servers should keep a record for your IP-address-to-machine-name mapping);
- the PC uses the IP address to retrieve the Web page from 'www.mynet.bt.co.uk'.

## 2.8 IP Addressing

Every machine on an IP network is identified by an IP address. In simple terms this may be thought of as the machine's telephone number. Each IP address consists of four numbers in the range 0 to 255, separated by a decimal point. e.g. '194.72.10.20.' All the machines on a given network will have IP addresses allocated from the same block, e.g. '194.72.10.20' and '194.72.10.21' would be on the same network.

The area of IP addressing is large and complex. The restrictions imposed by an addressing scheme devised with no knowledge of how large the Internet would become means that there are many modifications to the original IP addressing plans. For more detail on how IP addresses are allocated and used on the Internet, see Chapter 14.

## 2.9 Enhancing IP Services

While the fundamental IP network provides a basic transport mechanism to convey data between two points, additional services and features may be incorporated into the network to enhance the IP service provided to the customer. This section will describe some of the developments in IP technology.

- Virtual private network (VPN) technology

    Many customers require a private IP network for use within their business (a private IP network is referred to as an 'intranet'), but cannot afford to build their own network. VPN technology allows a single IP network (perhaps the Internet) to carry traffic for many customers at the same time, without allowing those customers to communicate directly with each other and risk compromising security. Sharing the network in this manner reduces costs, while maintaining security. Chapter 16 describes the IP technologies used to create VPNs.

- Quality of service (QoS) or differential service (Diffserv)

  'QoS' refers to the measured and proven capabilities of an IP network. A QoS-enabled network is a network that is able to conform to a pre-arranged set of performance characteristics (e.g. availability, throughput, delay) and prove that it is doing so. A network that supports 'Diffserv' is a network that can supply more than one level of QoS characteristics, either for different customers, different traffic types, or both.

- Multicast

  If a computer needs to send the same data to many other computers, there are three options:

  — send the data to each receiving computer, one at a time ('unicast');

  — broadcast the data to all computers in the network (very impracticable in a large network);

  — use multicast to send that data once, and have the IP network copy that data to all the computers that wish to receive it.

  To use multicast, the IP network needs to be enhanced with multicast-capable devices and the computers using the network need to be able to use multicast protocols.

- Caching

  When a user retrieves a page over the World Wide Web, the page is often transmitted from a great distance away, using expensive network resources on its journey. It is likely that the same Web page, or elements of it, has already been transmitted across the network in the recent past. Thus savings may be made in network resources and in the speed of retrieval if the network keeps copies of Web pages it has retrieved in a 'cache', then checks this cache when a new request is received to attempt to service all or part of the request. Networks may perform this task without the end users being aware (transparent caching) or allow the users to configure the caching feature for themselves by configuring their own browser to ask the cache for Web pages directly.

## 2.10   Internet Service Provision

An organisation that provides access to the Internet is called an Internet service provider (ISP). While providing access to the Internet, an ISP is in fact making up part of the Internet, since the Internet may be viewed as many ISPs, businesses and individual users all joined together and sharing traffic. The key elements of an ISP are summarised below.

22  *Conclusions*

- Access network

    This is an IP network which connects to customer sites using an array of different access technologies — leased lines, PSTN and ISDN dial, DSL. The access network collects IP traffic from all the customers of the ISP into IP PoPs (points of presence) and then delivers it to the core network.

- Core network

    This is usually a large, resilient network that is used to carry customer traffic across the network, either to another customer, or out of the network altogether to another ISP. Connections within the core network are referred to as trunk connections. Connections to other ISPs are referred to as transit connections. Chapters 3 and 4 describe how core networks are designed and the art and science of connecting to other ISPs .

- Server farm

    An ISP must provide many application services, ranging from the basic IP services such as DNS and mail, through to Web-hosting facilities. The servers used to provide such services are often grouped together in a few locations, called 'server farms'.

- Network management centre

    The network is administered from one or more locations referred to as network management centres (NMCs) or sometimes network operations centres (NOCs). The management centres deal with all aspects of running the network, from adding new customer connections, dealing with faults, to overseeing the installation of new capacity.

## 2.11  Conclusions

The machines, protocols and techniques used in IP networking are in a state of constant and accelerating change, driven by the exponential growth of the Internet. This chapter has aimed to describe why IP is so successful, what an IP network looks like and how it works — these fundamental concepts of IP networking are perhaps best summed up by returning to the 'post office' analogy.

A person (computer) wishing to send a parcel (IP packet) knows the name and address of the person to whom they wish to send it (hostname), and then checks to find the correct post code (DNS look-up to find IP address) which they write on the front of the parcel. The parcel is posted in a postbox (customer site access router) from where it is transferred to a van which runs on an A road (leased line) to the nearest post office (ISP's access router). The post office sorts the parcel based on the post code and decides how best to forward it (router performing routing look-up in a routing table) and sends it in a lorry along a motorway (SDH high-speed leased line) to a main sorting office (core router). The main sorting office sorts it down to a local office (access router) from where a postal worker delivers it to the final destination (access link). If the parcel was damaged or lost in transition, someone from the sending side would re-send a replacement parcel (TCP), unless it was junk mail, which would not be checked or re-sent (UDP).

Finally, the post office worker has witnessed an ever-changing array of new sorting machines, better lorries and vans, faster roads, more homes and businesses, road congestion, new road pricing schemes, huge amounts of post going overseas, new ways of using the post network and a rapid increase in the number of organisations offering postal service.

# 3

# SCALING GLOBAL INTERNET NETWORKS

## A Howcroft

## 3.1 Introduction

This chapter presents the technical details of how to design large IP networks. An *ad hoc* collection of routers and nodes grown organically could lead to an unstable and congested network. The principles presented here are therefore not optional for large network designs. Anyone planning to design a large IP network should read this chapter — those only interested in the physical topology of a scalable network may want to skip section 3.3.4 on 'Routing Design', while those with an interest in routing protocols should read section 3.3.4.

The design of large Internet protocol (IP) networks, such as a Global Internet backbone, involves many design parameters, e.g. the number of anticipated customers, connections to other service provider networks, anticipated traffic flows, routing implementation and many others. In all these areas scaling is vital to ensure the network not only works and is stable initially, but can grow to accommodate more customers, increased traffic and more points of connection.

The chapter uses the design of BT's and Concert's IP backbones, which form part of the global Internet, to demonstrate the techniques employed. Particular attention is given to the approaches used to scale the network while maintaining stability of the routing protocols employed.

## 3.2 Design Requirements

Many factors have to be considered in designing a large Internet network. An example of some of the typical requirements that have to be considered in the design process, broadly grouped into different categories, are listed below. Scaling has to be considered in all of the categories when designing the network.

- Customer access:
    — customer type, e.g. ISP, corporate, dial;
    — access types, e.g. leased line, dial, MAN;
    — access speeds, e.g. 64 kbit/s to 155 Mbit/s;
    — resilience, e.g. dual connection to the network.
- Node:
    — location, e.g. preferred cities/towns for customer and Internet access;
    — accommodation, e.g. space, power, air-conditioning;
    — network facilities, e.g. access for customer and core network circuits;
    — equipment, e.g. preferred network equipment types.
- Topology:
    — circuits, type of underlying network infrastructure, e.g. leased circuits, ATM;
    — resilience, network availability.
- Routing:
    — architecture, top-level network structure;
    — protocols, preferred routing protocols.

## 3.3  Basic Internet Design Techniques

Global Internet networks are designed around a number of nodes, also known as points of presence (PoPs), where customers and other networks are connected, joined together by wide area circuits (see Chapter 5). Nodes consist primarily of interconnected routers at a single location that are responsible for determining where individual IP packets of data should be forwarded. Each IP packet carries its own source and destination address and the router's forwarding decision is based on internal tables held within each router that determine to which of the router's various connections the packet should be sent, based on the destination address of the packet. The router's internal tables are dynamically constructed from information gained from the routing protocols that operate between the routers and are responsible for determining the preferred path through the network for each individual destination IP address.

To allow for network scaling, operational considerations and controlled connection to the rest of the global Internet, nodes may be grouped together into separately controlled segments known as domains or autonomous systems (ASs). Domains constrain the size and scope of the routing protocols employed and are a primary technique used in scaling the global network. Within each domain different routing designs and techniques may be used; the protocols used within the domain are known as interior gateway protocols. Standard interior gateway protocols are open shortest path first (OSPF) [1] and intermediate system—intermediate system (IS-IS). A common routing protocol is used to control inter-domain routing between interconnected domains of the global network and between each domain and the rest of the Internet; this protocol is known as an exterior gateway protocol. The standard exterior gateway protocol for the Internet is the border gateway protocol (BGP) [2]. Connection of two Internet network domains controlled by BGP routing is known as peering (see Chapter 4).

Interior and exterior routing protocols are designed to perform different tasks. Interior protocols look after choosing the best path between routers within the network based on the state of the network routers and links, and the costs given to the various alternative paths between interconnected routers. Exterior protocols look after choosing the best path to global Internet destinations based on the policies the various domains have regarding the use of their connection to other domains. Routing protocols are designed to take care of failures of individual routers and inter-router connections within a network. When a failure occurs the routing protocols operate to determine the best alternative paths available between routers and domains and rebuild the routing tables in individual routers accordingly. Interior protocols are optimised to rapidly react to failure and select an alternate path, while external protocols may be damped to minimise the propagation of route changes across the global Internet. Internet networks utilise BGP internally (iBGP) within a domain as well as for exterior domain routing (eBGP) as it is better suited than internal protocols such as OSPF for carrying the large number of Internet routes.

Individual routers within the network perform two main functions — packet forwarding based on a routing table, and route computing via the routing protocols which construct the routing table. Although these two functions are often taken care of by separate elements within typical routers in use within the Internet, the route computation process can be both time and CPU expensive. If the network is unstable the routing protocols may be unable to converge on an agreed view of the network; in such cases the network has effectively failed and packet forwarding, the main purpose of the routers, will be unable to function correctly. For these reasons network scaling and stability are among the main goals of global network design where the number of routers and possible routes between them within the network may number in the thousands. In addition to the interior routes the network has also to cope with 70 000 plus Internet routes and prevent any instability within the rest of the Internet affecting the internal network.

### 3.3.1 High-Level Network Architecture

The first task in designing a global Internet network is to determine its domain structure and the policy of each domain with respect to its directly connected domains within the global network and to the rest of the Internet. The number of routers and the interior routing protocol design within the domain determine the ultimate scale of each domain. Typically domains may span an individual country, a region such as North America or Europe, or be a global domain spanning many countries and regions.

There are three domain models prevalent in global Internet design as shown in Fig 3.1:

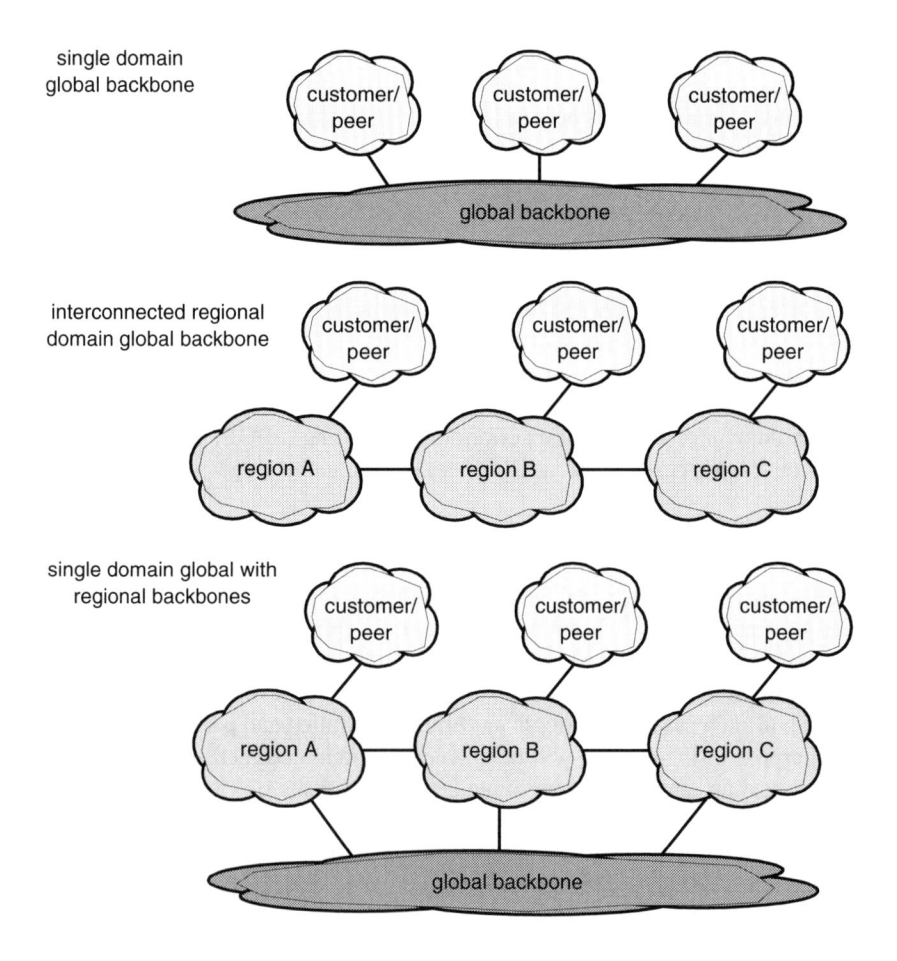

**Fig 3.1** Global domain structures.

- a global backbone configured as one autonomous system — this backbone will serve as the backbone for all regions it spans;
- a global backbone configured as a number of regional networks, based on geographical location (e.g. Europe, US and Asia/Pacific) each with their own AS — a regional network will peer with adjacent connected regional networks to form a combined global network;
- a global backbone configured as one AS — this backbone is used to interconnect a number of regional networks each with their own AS, while a regional network will peer with the global backbone to form the global network.

The following list gives determining factors in the choice of which of the above options are feasible:

- number of routers that participate in the interior routing protocol implementation;
- number of nodes;
- global or regionalised operational model;
- consistency of service;
- customer perception;
- routing complexity;
- mutual peering implementation;
- advanced feature implementation, e.g. QoS, VPN.

Of the top-level architecture models the single global domain network is only suitable for relatively small networks, in terms of numbers of nodes and routers, without very complex routing implementations, but does offer advantages in customer service, perception and advanced feature implementation [3].

The multi-domain architecture comprising interconnected regional domains offers scaling superior to a single domain structure and allows for regionalised routing policies. Its disadvantage is the increasing cost of domain interconnection required as the number of domains increases, unless the domains are simply connected as a chain, in which case Internet BGP routing is compromised for the ISP customer.

The global backbone interconnecting separate regional domains is a fusion of the two other models and overcomes some of the disadvantages of both at some additional cost. It also offers the benefit of global as well as regional peering policy implementation; the Concert IP backbone follows this design.

### 3.3.2 High-Level Network Design and Topology

Once the domain structure of the global network is decided, a top-level network architecture/design and topology needs to be defined. Here more detailed requirements such as node locations, customers and traffic forecasts, performance and costs need to be considered. Equally important, and sometimes overlooked, are the IP routing requirements to be able to route traffic as desired between the various ingress and egress points on the network via the desired traffic paths.

In general, whatever type of domain is being considered, the network hardware and logical design should be broken down into different layers to provide scalability and stability. In global design, this methodology results in a hierarchical or tiered design which may be applied to any of the domain options chosen as described below:

- a tier-1 or global core network joining major large nodes forming an inter-continental backbone;
- a tier-2 or regional core network interconnecting large/medium nodes forming an intra-continental backbone connecting into the tier-1 network;
- a tier-3 network interconnecting small/medium distribution nodes typically within countries connecting into the tier-1/2 network.

Each of the network tiers serves two functions:

- concentration and localisation of traffic within a tier;
- segmentation of the network at the routing level.

Tier 1 and tier 2 may be combined into a single tier for routing purposes within regional domains as this more closely matches the capabilities of IGP implementations. The tiered network layers are illustrated in Fig 3.2.

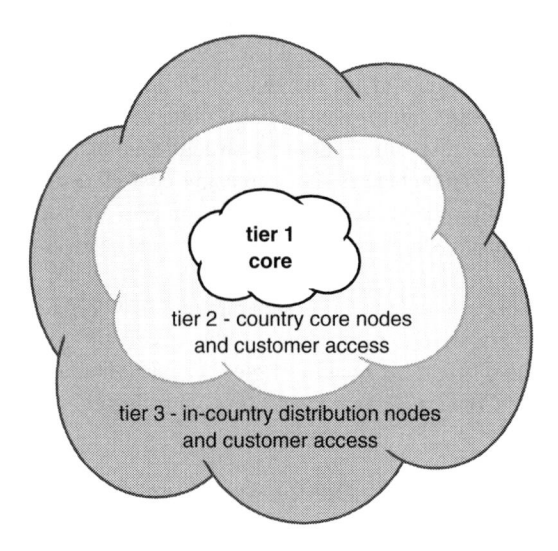

**Fig 3.2** Tiered network hierarchy.

Topologically the tier-1 and tier-2 networks typically comprise a mesh of high-capacity wide-area circuits connecting major nodes. These nodes serve as interconnection points to other global or regional ISP networks forming the Internet, and concentrate customer traffic from the tier-3 distribution network as shown in Fig 3.3.

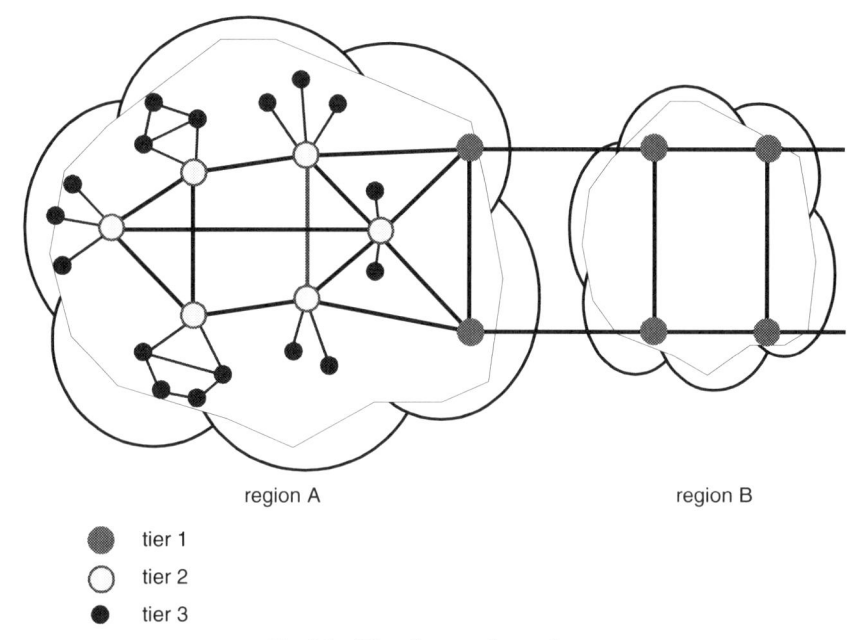

**Fig 3.3** Tiered network topology.

region A    region B

● tier 1
○ tier 2
● tier 3

### 3.3.3  Node Design

Node design is primarily concerned with the interconnection of routers in a single location. Ideally a single router could fulfil the requirements of customer access to the network and the interconnection of core network circuits. Unfortunately current router implementations do not provide a device that can scale to meet the conflicting demands of both access and core and it is often necessary to separate these functions into different physical devices.

Customer access routers should provide for all the different access types required in the same physical router and provide the maximum flexibility in choosing the mixture of access interfaces to be deployed. Ideally the access router should permit as many customers to be connected to a single router as possible, through either multichannel interfaces connected to layer-1 channelised services or interfaces connected to layer-2 access networks such as FR or ATM. These permit individual customer access connections to be aggregated together by the access provider, typically a PTT, and be delivered to the node as a broadband interface carrying many different customer accesses. To allow the access router to be connected to other routers and node infrastructure high-speed LAN and wide-area point-to-point interfaces must also be provided. Apart from the obvious routing functions, the access router must be capable of supporting 'edge-of-network' functions such as eBGB routing and filtering, traffic policing and quality of service (QoS) implementation to/from the customer.

Core routers should be able to route IP packets as fast as possible with minimum delays across a selection of very high speed local and wide-area interfaces. Typically in today's networks, gigabit and terabit switching/routing capacity in core routers is required. Core routers will also be required to provide for different packet queuing mechanisms on physical interfaces that enable QoS schemes to be implemented on the core network. Access and core routers must be interconnected together in a cost- effective, scalable fashion that permits expansion of customer and core network in both number and scale of circuits. Another key goal is resilience in the network core so that any core circuit or router must be able to fail without affecting node functionality. The scaling requirements on the different tiers of nodes may be different and node design and routers used may be changed accordingly. A limit on the scalability of the node is primarily the switching capability of a single router, the number of physical interfaces the router can support, and the speed of interfaces available. If the node is interconnected by point-to-point interfaces, the ultimate scaling of the node is typically constrained by the number of interfaces supported, whereas if the node is interconnected by high-speed LAN, node scaling is typically limited by LAN bandwidth. These requirements lead to a node design of two or more core routers, interconnected to a number of access routers, interconnected by a number of independent Gigabit Ethernet LANs, supported by high-bandwidth, expandable LAN switches. Routers are connected to at least two physical LAN switches by different interfaces for resilience and load scaling. This node design has been used very successfully throughout BT's and Concert's high-speed backbones and continues to be the preferred design. Figure 3.4 shows a typical tier-2 node design.

### 3.3.4 Routing Design

Routing design is key to the stability and scalability of the network. Routing also has constraints in its ability to send packets along the preferred paths; for these reasons routing design must always be considered before the physical design of the network as some physical implementations will not meet the design criteria [4].

Scaling Global Internet Networks  33

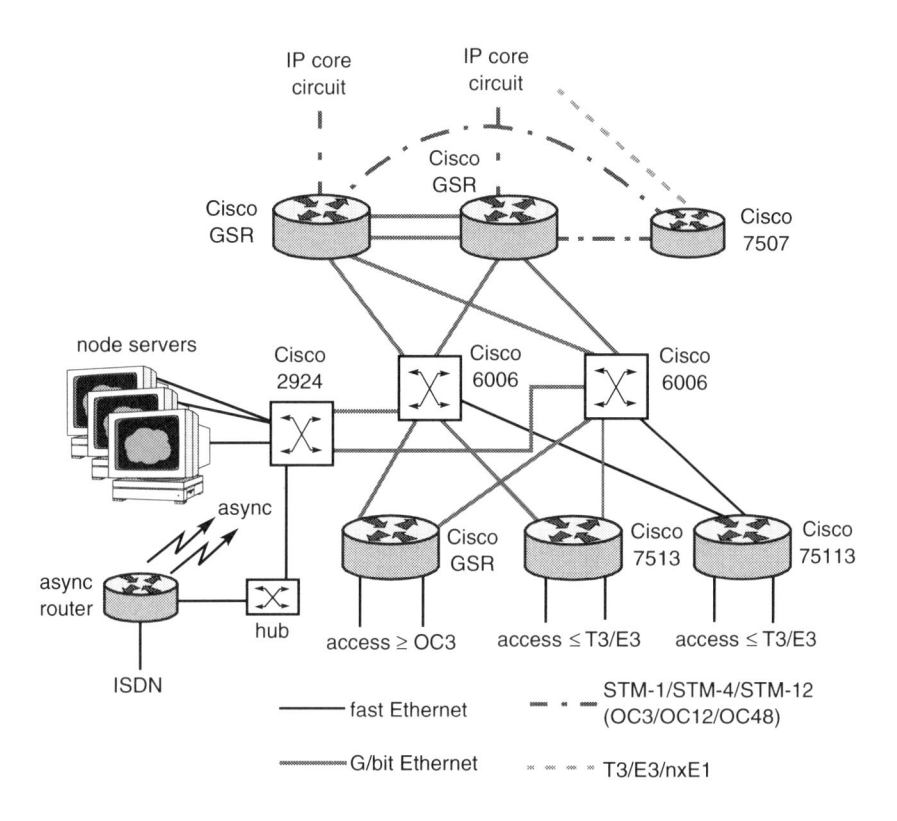

**Fig 3.4** Typical tier-2 node design.

### 3.3.4.1 Principles

The goals of routing design are stability, resilience, speedy response to change, route integrity and manageability, while permitting the network to scale to meet the foreseeable customer growth and geographic reach. Unfortunately these goals are, in many cases, not complementary and compromise is required. A tiered network architecture as described earlier allows routing to be compartmentalised within different domains and tiers. Within each domain an IGP and iBGP will be used; in general the IGP will be used solely to look after the internal state of the domain while iBGP will be used to carry the state of external routes. By doing this, the IGP can be tuned to provide rapid response to link/router failure within the domain, while isolating the BGP from many of these changes.

### 3.3.4.2 Interior Gateway Protocol (IGP)

The primary choice of IGP is between IS-IS and OSPF and both of these protocols are in wide use throughout the Internet. Until recently IS-IS had been the protocol deployed in most large IP backbones, although this choice was primarily due to it being more stable than OSPF on Cisco routers. OSPF has begun to be used by more networks recently as it offers some benefits over IS-IS in its multi-area support and is available on a wider range of different vendor routers.

Both IS-IS and OSPF protocols are capable of scaling to similar numbers of link state advertisements (LSAs). LSAs are the database elements that are exchanged between routers by the routing protocol. They contain information regarding the state of the adjacencies between the routers and IP addresses, within the domain, that are reachable via that router. LSAs are exchanged to signal the state of links joining the routers together; any link-state change is distributed to all routers within the IGP which then have to recompute their view of the network. The process by which LSAs are spread to all routers within the IGP domain is almost an $n^2$ process, as each router forwards a newly received LSA to each of its directly connected neighbours, except the one from which it received the LSA. Thus, the LSA process floods the network with IGP protocol information. This takes away bandwidth available to user data and utilises CPU power at each router to ensure all its neighbours have up-to-date copies of the LSA.

Once LSAs have been flooded across the network, all routers compute their one routing table using the Dijkstra shortest path first (SPF) algorithm. This is computationally expensive on the router CPU — the larger the network the greater the number of LSAs and the longer the router takes to recompute its routing table.

The routes or LSAs that the IGP should carry must be kept to a minimum to avoid the $n^2$ problem, as must the frequency of the LSAs. For this reason it is sensible to carry customer access link information in the iBGP rather than the IGP. Good infrastructure IP addressing schemes, that hierarchically address the network components and allow address aggregation, are essential. Typically IP addressing schemes assign blocks of IP addresses to specific nodes so that, at the core routing level, a node may be seen as one aggregate IP address rather than multiple individual addresses (see Chapter 14).

The IGP also sends periodic 'hello' messages between neighbour routers to determine link integrity; if a link failure is detected, the LSA flood process and SPF computation are triggered. The frequency of hello messages is a trade-off between good routing convergence and network stability. In a network with highly stable wide-area links, such as those derived from underlying SDH implementations, it is possible to reduce the frequency of periodic hello messages.

As network bandwidths become much larger and routers employ much more powerful CPUs, it is becoming possible to safely increase the number of routers within a single IGP area. BT and Concert have chosen OSPF for the IGP implementation. Concert decided that a reasonably safe limit for OSPF scaling was a few thousand link-state advertisements, typically equating to several hundred routers. There are other major Internet backbones using IS-IS as their IGP that have been successfully scaled to around one thousand routers.

### 3.3.4.3 Border Gateway Protocol (BGP)

It is useful to look at the design of iBGP and eBGP separately as each has different goals and design principles. However, both implementations of the protocol are used to carry the full Internet routing table in major IP networks and one of the main goals of BGP design is to limit the susceptibility of the network to changes in the global Internet routing table.

As described previously eBGP is concerned with the control of routing between separate ASs within the Internet. For a typical global IP network eBGP will be used to control routing between its own ASs and those of its peer connections to the rest of the Internet. eBGP is also used for control of customer routing when the customer network is connected to more than one AS within the Internet, and is used to announce and receive selected routes between peering ASs. The network receiving routes from a peer will only be able to send traffic to those routes received, assuming no default routing is employed (default routing on eBGP peering is strictly forbidden within the Internet although it is difficult to detect). It is also possible to filter routes received via eBGP so as not to accept those not required. A network's routing policy describes its implementation of eBGP route advertisement and acceptance and is based primarily on the peering agreement between the connected networks. For manageability and scalability it is useful to have an automated system that implements the network's routing policy, deriving eBGP filter configurations from a high-level description of the policy.

It is good practice, to help maintain network stability, when eBGP peering with other ASs, to implicitly deny any route announcement and then build route filters that explicitly allow only those eBGP route announcements from peers that the routing policy permits. It is also good practice to only accept route announcements from ASs that are allowed to send them.

When receiving eBGP routes all routes are initially sent on eBGP initialisation between neighbours; route changes are then exchanged as route reachability changes. If a route that is being received is unstable, known as route flapping, this will cause excessive BGP process computation on the affected router(s) and also consume bandwidth on the peering link. It is useful to employ BGP route damping, which removes an unstable route until it becomes stable, at the periphery of the network. To help protect other networks against unstable routes and to minimise the size of the Internet routing table, internal network routes, including static customer routes, should always be announced as the largest possible aggregate.

The number of eBGP peerings and alternative paths has a direct cumulative effect on router memory and CPU utilisation. For this reason it is recommended to reduce alternative paths to the minimum required for resilience.

Within an AS, iBGP is used to carry the Internet routing table as received from eBGP. To avoid routing loops all routers in the network must maintain iBGP connections to all other iBGP routers; this results in another practical scaling limit as each router has to establish $n-1$ iBGP sessions. This again has a direct relationship on router memory and CPU utilisation. A practical upper limit of approximately 100 BGP sessions per router is considered safe.

## 3.4 Network Scaling

As the network grows to meet the ongoing demands of geographical expansion, increased customer connections at each node and growth in customer bandwidth, then more nodes, additional access router interfaces and increased core bandwidth will be required. Within certain limits the network design principles outlined earlier should be able to scale sufficiently provided hierarchical design has been used. However, once an individual routing domain includes several hundred routers and a complicated core mesh of wide-area circuits, a simple single-area IGP implementation and a full-mesh iBGP will be nearing the limits of stability and scalability.

To enable a single routing domain to scale significantly above the few hundreds of routers, IGP routing will need to make use of the multi-area capability of OSPF (IS-IS multi-area routing implementations have had restrictions caused by default routing between areas that make it unsuitable for Internet backbones, although there is now a draft standard on domain-wide prefix exchange for IS-IS). In addition, the iBGP implementation should make use of either route reflectors or BGP confederations to reduce the scale of the internal iBGP mesh. For these techniques to be employed effectively the network should initially be designed with them in mind to provide the required architecture, IP addressing and resilience.

Node design will also need to be considered to allow the connection of many routers, particularly at the tier-1 and tier-2 nodes. Intra-node connectivity must provide sufficient bandwidth within the constraints of router chassis expansion capabilities, not to mention cost.

### 3.4.1 Node Design

As the number of routers at a node increases, the LAN infrastructure needs to accommodate the growth in number of LAN connections and the intra-node bandwidth. Provided LAN switches with sufficient interface and bandwidth capacity have been used, growth is relatively easy up to the limiting point at which the LAN bandwidth to/from the core routers approaches the practical LAN limit. Typically it is wise not to exceed 70% average loading on a switched Ethernet connection.

To scale to greater intra-node bandwidth, additional LAN segments may be deployed to create more parallel paths between routers. If this does not suffice, selective intra-node point-to-point links or a higher speed LAN technology, such as the emerging 10 Gigabit Ethernet, will be required. In practice, the core router switching and interface capability will often be exhausted before the intra-node LANs are full. New enhanced switching capacity core routers will be required at this time and can be phased into the design while utilising the old core routers as very high speed access routers.

### 3.4.2  Routing Domains

It is always possible to expand the network by creating additional routing domains. This does impose some limitations on the top-level architecture and traffic paths that may be unacceptable. Certainly when considered from the customer point of view, too many BGP domains in a single-provider network is not viable as BGP customer routing and associated traffic paths become difficult to control. The number of ASs between a customer and a potential source/destination network is crucial to ISP and content provider networks as, in general, the preferred route is through the lowest number of AS 'hops'.

### 3.4.3  IGP

Both OSPF and IS-IS may be configured into a multi-area implementation of a backbone area connecting a number of leaf areas together. At the boundary between the backbone and the leaf area it is possible to isolate LSAs from one area entering the other. As mentioned previously, IS-IS currently has a restriction in its multi-area implementation that means only default routes may be advertised from the backbone area to a leaf area. This creates problems in handling route selection for traffic engineering and IGP-derived BGP metrics, which makes it unsuitable for complex Internet backbone networks. OSPF also has restrictions in that a leaf area may only be connected to the backbone area. However, it is possible to make all access routers at a node part of a leaf area or, a number of tier-3 nodes may be homed to a single backbone tier-1 or tier-2 core node, for example. Careful IP addressing design is required in multi-area planning to maximise its scalability in terms of route summarisation from leaf to backbone area. There are also useful benefits in terms of routing isolation between backbone and leaf areas and the stability this may provide to the majority of the network should a leaf area routing become unstable even without route summarisation. OSPF multi-area implementation is most commonly

38  *Network Scaling*

achieved by designating the core routers at a tier-1 or tier-2 core site as the area border routers (ABRs). This does impose the restriction that the core routers must have a routing connection in the backbone area as well as in the leaf area. This may be achieved by a separate physical connection in both the backbone and leaf areas or by dividing core router intra-node LAN connections into separate virtual interfaces using protocols such as IEEE802.1q [5].

A multi-domain, multi-area OSPF routing scheme is illustrated in Fig 3.5.

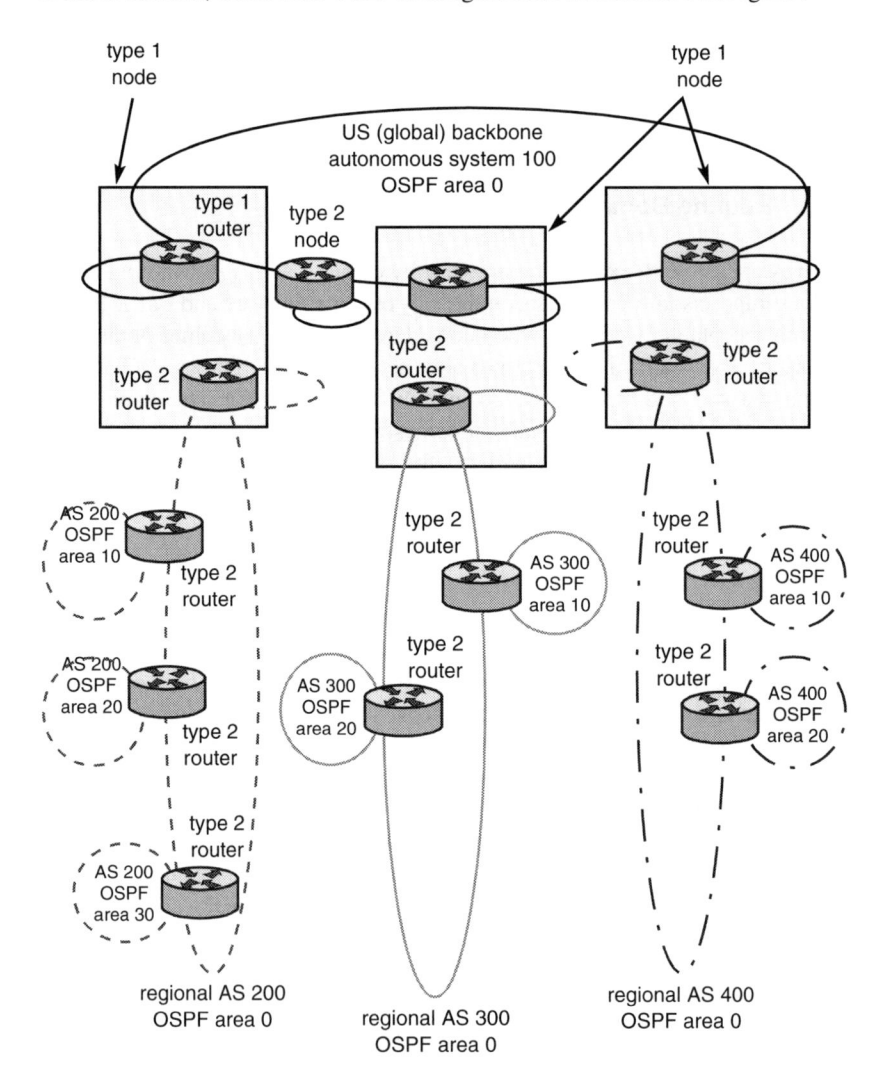

**Fig 3.5**  A multi-domain, multi-area routing scheme.

### 3.4.4 BGP

There are two primary iBGP scaling mechanisms — route reflection and confederations. Within the Internet, route reflection is the most commonly used technique. Route reflectors are designated routers that may serve a number of route reflector clients where the clients do not need to be in the full iBGP mesh to gain access to full routing information. For correct route reflector implementation providing routing resilience, the route reflectors must be fully meshed and their implementation should match the physical resilience of the network to provide routing resilience in case of network failure. Typically two core routers at a node are designated as route reflectors for that node and all the other node routers are designated as route reflector clients. Provided the core routers are connected in multiples to other node core routers and to their respective route reflector clients, then failure of any single core router at a node may be tolerated. Network scaling is improved with route reflectors because the route reflector clients are no longer in the full iBGP mesh. The route reflectors themselves provide the limit of scaling as they must be fully meshed and have individual iBGP connection to their respective clients. Taking the 100 iBGP sessions per router as a guide and two route reflectors per node it is possible to have 80 nodes with each node having 20 access routers.

With very careful design and matching the physical network to the iBGP implementation, it is possible to have a two-level route reflector hierarchy which would improve iBGP scaling even further, although in practice this is not currently required for present network sizes.

BGP confederations allow a single BGP AS, as seen by its eBGP connections, to be constructed of multiple ASs. This allows the multi-domain scaling approach to be used as any particular domain gets too large without the customer BGP restrictions mentioned earlier. Confederations do have a drawback though, in that traffic will always take the internal route within an AS, even though it is confederated with other ASs. This may lead to traffic flowing across core circuits different from those intended, unless the network topology and desired traffic flows are closely matched to the internal confederated domain structure. The implementation of BGP confederations together with route reflectors may extend the scalability of a single domain even further but the feasibility and design principles have not been established, to the author's knowledge. BT and Concert have employed route reflector BGP implementations for their IP backbones, as have most Internet backbone providers.

### 3.4.5 Traffic Scaling

As the network grows, the volume of traffic will inevitably increase. Internet traffic has experienced growth rates of 400% per year over the last five years and shows no sign of decreasing. Such growth rates demand a scalable backbone with flexible, well-planned means of core network bandwidth expansion.

Within the Internet it is inevitable that the majority of traffic flows will be off the network, as a single IP backbone, however large, will constitute only a small portion of the global Internet. It is therefore vitally important to ensure that sufficient interconnectivity to other Internet networks is provided at all network tiers to ensure that traffic is not carried unnecessarily across the core network. Peering with other major providers at the country, regional and global level at all major node sites ensures that the majority of off-net traffic is carried for the least distance on the core network. However, multipoint peering with other backbones at the regional and global level is arranged so that each peering network sends its traffic destined for the other network to the nearest peering point; this process is commonly referred to as 'hot potato' and is achieved by allowing the IGP to decide the nearest exit. When peering with content provider networks it is important to ensure that the content network delivers the traffic at the nearest peering point to the node for which it is destined. This is achieved by sending multi-exit discriminators (MEDs) to the peer network to let it know to which peering point to deliver its traffic. Major peering locations should therefore be located on the main core network at tier-1 or tier-2 node sites.

Core routers must be capable of coping with at least one to two years anticipated core network bandwidth growth requirements. This demands sufficient chassis capacity for additional core network interfaces in order to add new core wide-area circuits. Additional core capacity may either be higher capacity circuits to replace existing circuits or additional circuits, either supplementing existing circuits or adding new paths to the network topology where required.

To enable the network to grow successfully it is essential that network traffic flows are measured and their growth trends understood. This has traditionally been very difficult on IP networks but recent developments, such as Cisco's Netflow, provide the means to determine traffic flows and should be deployed as an integral part of the network.

### 3.4.6 Traffic Engineering

It is often necessary to route certain traffic across network routes that would not normally be used, as determined by the routing protocols, to relieve temporary network hot-spots or to make use of currently underutilised network paths. It is possible to achieve this on a macro-scale by adjustment of IGP routing metrics, but this may also lead to undesirable traffic flows as well as being difficult to manage. Individual traffic flows may be adjusted by the use of BGP techniques such as MEDs for balancing traffic on multiple eBGP connections to other networks and the tagging of customer and peering routes with different BGP community values to allow specific route announcement filtering. A relatively new protocol, multiprotocol label switching (MPLS), offers the possibility of enhanced traffic engineering and increased network scalability by assigning labels to IP packets associated with particular network flows at the edge of an MPLS-enabled network. The labels are used by MPLS-enabled routers within the core to switch the packet along a predetermined label switched path (LSP). Individual LSPs between router pairs may take different paths through the network allowing traffic flows to be adjusted as required. If the core routers of a conventional routed network, as described previously, were MPLS-enabled, the core routers would no longer need to participate in the full network routing protocols, thus enhancing scalability (see Chapter 15).

## 3.5 Conclusions

This chapter has highlighted major factors that affect network stability, particularly routing design and implementation. The principles and techniques used by Concert and BT to scale IP backbones to provide stability and potential for growth have also been outlined.

New protocol developments, such as MPLS, improved multi-area IGPs and router implementations that decouple the packet forwarding and routing functions, will also aid in network scaling.

The challenges for IP network designers are increasing all the time with the demand for multicast, differentiated services and IPv6 likely to increase — all of these will increase network complexity and consequently affect network scalability and stability. Even in the face of these new challenges the basic IP network design principles of hierarchical design and routing segmentation controlled by a manageable routing policy will still apply.

## References

1  Moy, J.: '*OSPF Version 2*', RFC 2328 (April 1998).

2  Rechter, Y. and Li, T.: '*A border gateway protocol 4 (BGP4)*', RFC 1771 (March 1995).

3  Whalley, G. et al: '*Advanced internetwork design — today and in the future*', BT Technol J, **16**(1), pp 25-38 (January 1998).

4  Yu, J.: 'Scalable routing design principles', UUNET (April 1999).

5  IEEE802.1q Standard: '*Local and metropolitan networks — virtual bridge local area networks*', — http://standards.ieee.org/catalog/IEEE802.1.html

# 4

# THE ART OF PEERING

## S Bartholomew

## 4.1 What is 'Peering' — an Introduction

This chapter looks at the essential ingredient of the Internet — how the many different Internet service providers connect to each other so that any*one* on the Internet can reach any*thing* on the Internet. The general term used to describe how ISPs connect to each other is 'peering'. The implementation of an ISP's peerings will be a major determinant of the performance of the services the ISP offers to its customers. This chapter is an introduction to both the key principles and the art of peering.

Peering is simply defined as an interconnection between two networks which facilitates the inter-exchange of data, and has become one of the most important and most widely debated subjects within the Internet industry. For an Internet service provider (ISP), it has a direct impact on the cost base, revenues, market position, customer perception, and network performance. There is no doubt that, as the Internet grows and the volume of traffic increases, the complexity of interconnections between the networks will increase. From an engineering perspective 'peering' is used as a generic description to cover an interconnection between two Internet networks. However, other terminology must be employed to describe the commercial framework that underpins such interconnects. Although many of the terms are loosely exchanged and change frequently, the following will be used for consistency in this chapter.

- Interconnection

  This is a physical and direct interconnection, a circuit, between two networks which facilitates the inter-exchange of data — the term is used without reference to the commercial agreement between both parties. Peering cannot occur unless there is some physical interconnect between ISPs.

- Peering

  Peering is the interconnection of two ISPs so that customers of one ISP can exchange traffic with customers of a second ISP directly, without transiting a third (usually higher order) ISP. The key words in the definition are 'exchange' and 'without transiting a third (usually higher order) ISP', as these describe an important concept, i.e. true peering means that the relationship between ISP peers is not one of customer/client. The peering interconnection between two carriers is based on a commercial agreement where each network exchanges traffic to and from their respective paying customers. Although discussed in a later section, the typical commercial scenario supporting this type of interconnect is based on a (50%—50%) shared cost model with no settlement for the volume of traffic exchanged.

Figure 4.1 shows two simple networks implementing a peering relationship. Rednet will advertise their paying customer's addresses over the interconnect to Greennet. These address announcements are often referred to as 'routes' or 'prefixes'. For simplicity they will be called RED ROUTES. In return, Greennet will advertise GREEN ROUTES to Rednet. The routers at both ends of the connection will now see an available path to the distant customer networks over the interconnect which can be used to exchange traffic. In order to get traffic flowing, three actions need to take place. Both networks must 'announce' routes originating within their network. Both networks will in turn 'export' the availability of the routes to the other network over the interconnection and in turn the other network must 'import' or accept the route as being reachable via this interconnect.

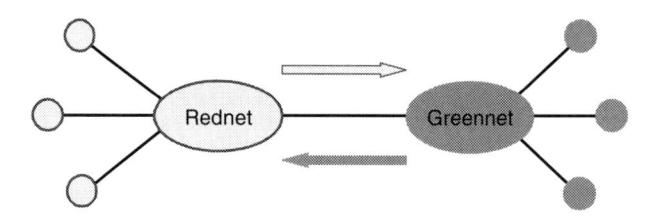

**Fig 4.1**   Simple peering between two networks.

In Fig 4.2 it can be seen that Greennet has now implemented a second peering with another network, Bluenet, and routes are being exchanged. Although there is a physical path between all three networks, traffic between Rednet and Bluenet will not be exchanged. This is because Greennet will not announce routes learned via one peer to another peer (a basic rule of Internet peering). As the Internet is a network where all networks must be reachable, clearly this is an unacceptable situation and an alternative solution must be found.

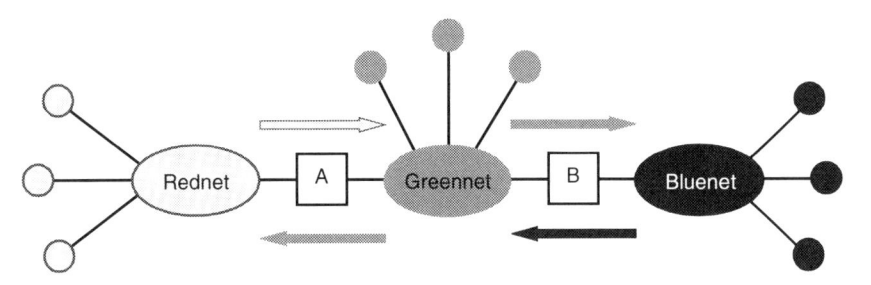

**Fig 4.2** Two networks peering with Greennet.

- Transit

    One way of overcoming the problem in Fig 4.2 is to route traffic via a third party network. Such networks are called transit networks and the product or service they provide is known as 'transit'. Unlike 'peering', the principal commercial model behind this type of interconnect assumes that a connecting network will pay the transit operator/provider to export its routes to other directly connected networks. They are also 'paying' to see all the transit providers' routes (both its customers and those learned via peering connections). The connecting network will also incur all associated costs of the interconnection, essentially becoming a customer of the transit provider. In Fig 4.3 Yellownet provides a transit service to both Rednet and Bluenet.

    The interconnection and commercial relationship between Yellownet and Greennet now becomes interesting — clearly there needs to be a direct interconnect, as neither Rednet nor Bluenet provides transit services. It could in fact follow either a transit or peering model — both would allow all the connected networks to pass traffic between them. It should be noted that if Yellownet and Greennet were directly connected, both Rednet and Bluenet could get access to Greennet over their respective transit interconnects as well as their peering interconnects.

    At this point the concept of an autonomous system (AS) must be introduced. An AS can be considered as one or a group of routers, that share a common routing policy and network management, but which are presented to external networks as one entity. Routing of traffic between these networks is achieved by using the border gateway protocol (BGP), which is the *de facto* inter-domain routing protocol presently used on the Internet. An ISP operating a network will have an AS number assigned by one of the regional Internet registries (see Chapter 14); these numbers are primarily used by the BGP to assist in routing decisions but are also used to understand ownership and determine network boundaries.

46  *What is 'Peering' — an Introduction*

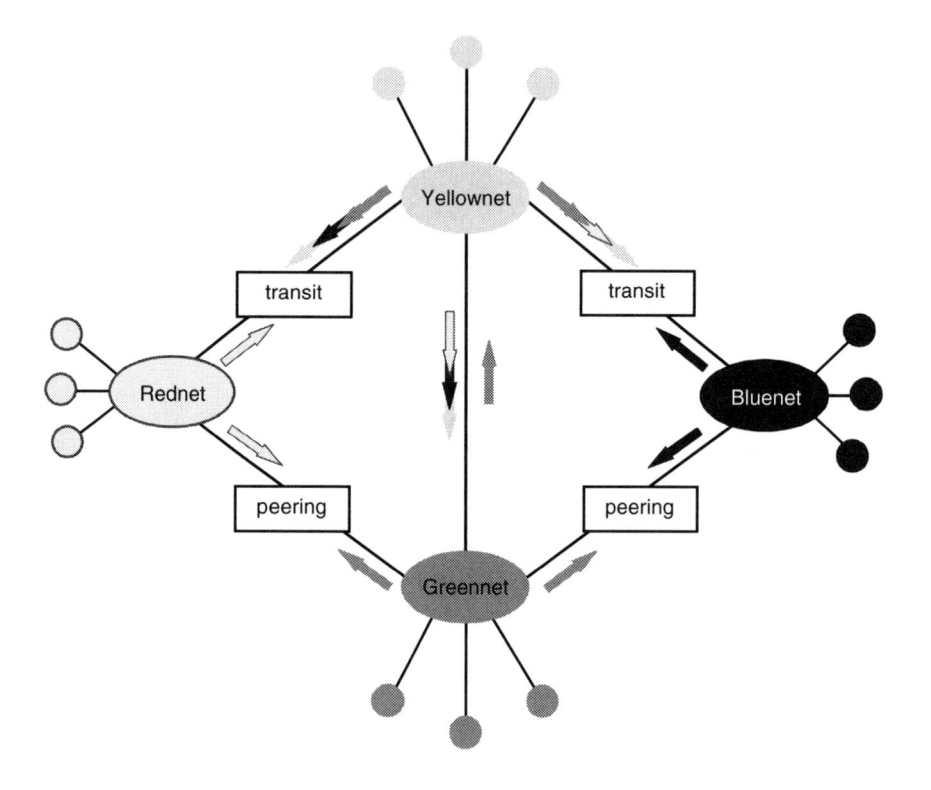

**Fig 4.3**   Yellownet provides transit to Rednet and Bluenet.

As a route is advertised each ISP network it traverses 'adds' its AS number to the route. When an ISP accepts a route into its routing table it will therefore not only have the destination but also the number of ASs (networks) the route has traversed. In cases where a destination network is seen via two or more interconnections, the BGP chooses the path with the least number of AS network hops to determine the routing path. In the example in Fig 4.3, all the networks have AS numbers assigned. Rednet would send packets to Greennet over the peering link, hop count = 1, rather than over the transit link via Yellownet, hop count = 2. Rednet would most probably wish to use this route from a purely cost basis.

As a general rule, an ISP will typically offer their directly connected (revenue generating) customers access to routes seen via their transit interconnects and their peering interconnects. An ISP will typically offer peers only routes to the ISP's directly connected (revenue-generating) customers — they do not offer peers routes to their transit interconnects and other peering interconnects, i.e. free transit.

## 4.2 Tier-1 Carriers

Returning to the above model, Rednet and Bluenet have grown and have discovered that they are exchanging large volumes of traffic between them and they are sending the traffic via Yellownet, a facility for which they are both paying. They decide to interconnect their respective networks on a 'peering' basis. Yellownet reluctantly, as they have now lost two revenue streams, agrees to peer with both these networks — this gives a fully meshed network 'peering' on a shared-cost basis.

In recent years the industry has adopted the term tier-1 carrier to describe networks that fit this model. To understand how they got there we need to look back to the Internet's formative years. In the USA in the early 1990s the National Science Foundation (NSF) operated and provided networking to the research and education community. The NSF established a new 'high-speed' T3 45-Mbit/s backbone, called the NREN (National Research and Education Network) with the specific purpose of conducting high-speed networking research. Although it was not to be used as a commercial network, it did, however, grant access to commercial networks under an 'acceptable use policy' (AUP). The AUP allowed these 'commercial' operators to pass traffic to the NSF data and computing centres, but did not allow commercial traffic to be exchanged between these commercial carriers over the NSF backbone. A number of network operators had a direct involvement in the operation and management of the NSF backbone — ANS, Cerfnet, Alternet, BBN, Sprint, and MCI. To comply with the AUP, but meet the emerging demand for commercial traffic exchange, these networks started to interconnect and 'peer', passing commercial traffic over those links outside the NSF backbone. As the Internet started to expand, new start-up service providers and large organisations saw the commercial benefits of access to this new network; the only option was to connect to one of the established but well-connected commercial networks. At this time BT formed the Concert alliance with MCI, and it was therefore natural for Concert and BT to obtain its transit services from the tier-1 provider (MCI). Concert and BT offered transit services by onward selling of the transit service they obtained from MCI.

This has determined the hierarchical US-centric topology of the present Internet. The label 'tier-1 carrier' was given to those networks who advertise all global routes and who interconnect with all their peers on a 'no-settlement, shared-cost' basis. Figure 4.4 demonstrates the relationship between tier-1 carriers using an example with four peers only.

## 4.3 To Transit or Peer

Returning to the US tier-1 model example with four networks offering transit services (Fig 4.4), the decision faced by two imaginary UK start-up ISPs will now be considered to illustrate the nature of the decisions Concert and BT have made for

## 48 To Transit or Peer

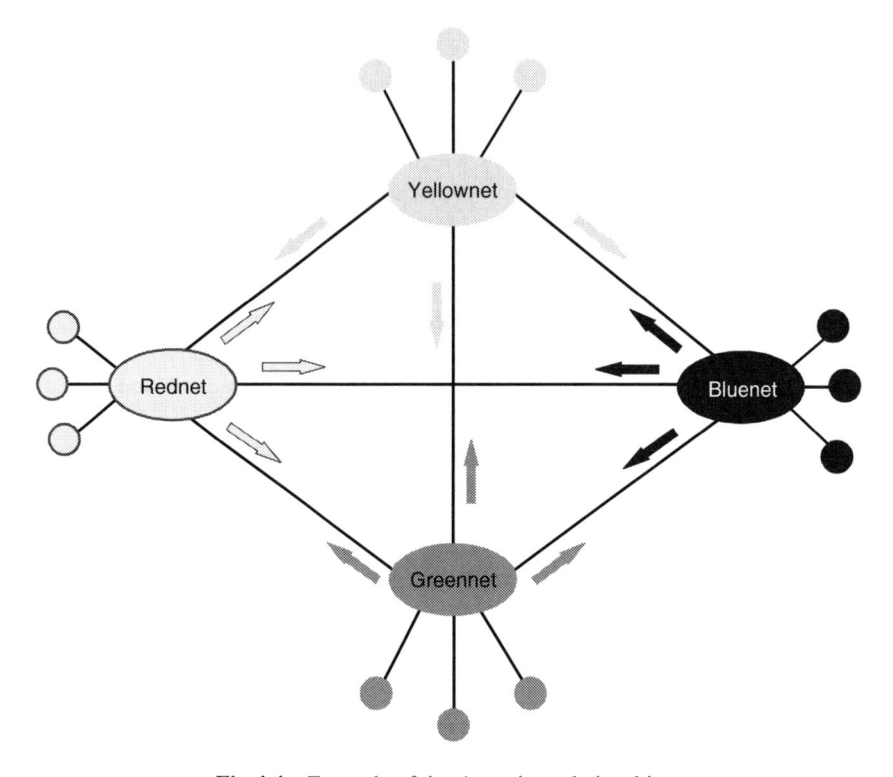

**Fig 4.4** Example of tier-1 peering relationships.

their UK IP service. Do they try to 'peer' with all of the networks or do they select one of them as a 'transit' to give global routes? It should be noted that, in the current Internet, the tier-1 networks may decide new entrants do not meet their criteria for peering and therefore, in practice, the question 'to transit or peer' may be academic — see section 4.6. From a purely cost perspective this can be shown as follows.

- To get all routes via transit provider can be expressed as:

    TransitCost = CostOfPort + InternationalCircuit

    Example transit port costs from a tier-1 US network at T3 speeds (45 Mbit/s) connection are between $35k—$80k per month (assumed at $50k per month), an international circuit UK—USA T3 (assumed at $100k per month); it is also assumed that the cost of the international circuit includes the 'land tails' to connect the international circuit to the PoPs.

    The transit cost per Mbit/s shipped per month (CPMS) is $3.3k. Note that these costs are indicative costs for this example only, since, as the sizes of the connections across the Atlantic rapidly increase, the costs also change dramatically.

## The Art of Peering 49

To get all routes via peering can be expressed as:

$$\text{PeeringCost} = N \times \frac{\text{CostOfInterconnect}}{2}$$

where $N$ is the number of networks required to get all routes.

Four international circuits UK—USA T3, assumed at $100k per month peering cost per Mbit/s shipped per month, is $4.4k. Note that many USA-based tier-1 network providers would not pay half the cost of the transatlantic circuit, thereby doubling the cost of peering.

The cost per Mbit/s shipped figure is often used as a benchmark to determine interconnection strategy.

Although bandwidth and port prices have fallen dramatically over recent years, purchasing 'transit' continues to be the least-cost option over full peering. It must also be noted it is unlikely that a USA-based tier-1 carrier would enter into a 'peering' arrangement with a new market entrant.

Having decided that a transit purchase provides the least-cost option, other factors will influence the choice of carrier:

- performance of the transit provider's backbone;
- location of access nodes;
- number of directly connected customers;
- market position.

In the case of Concert and BT, the decision is also influenced by its partnership with AT&T, which has its own USA-based network with tier-1 peerings. Similarly other ISPs may find their choice of peerings influenced by partnerships or joint ventures.

## 4.4 Peering Dynamics

In Fig 4.5 the model shows that the two UK ISPs (NetA, NetB) have chosen Rednet and Bluenet respectively as the preferred transit providers. They now see global routes via their transit arrangements. Any traffic between NetA and NetB will route via the USA.

In the same way, Rednet and Bluenet found that they were exchanging large volumes of traffic between them and moved to a peering relationship. Both NetA and NetB at some point will wish to interconnect directly to improve the performance between their customers. Typical round trip delay to the USA is around 100 ms. This could be reduced to less than 10 ms if the traffic stayed within the UK. A direct 'interconnect' between the networks would also significantly reduce the CPMS figure.

The following sections describe the ways of achieving this.

50  *Peering Dynamics*

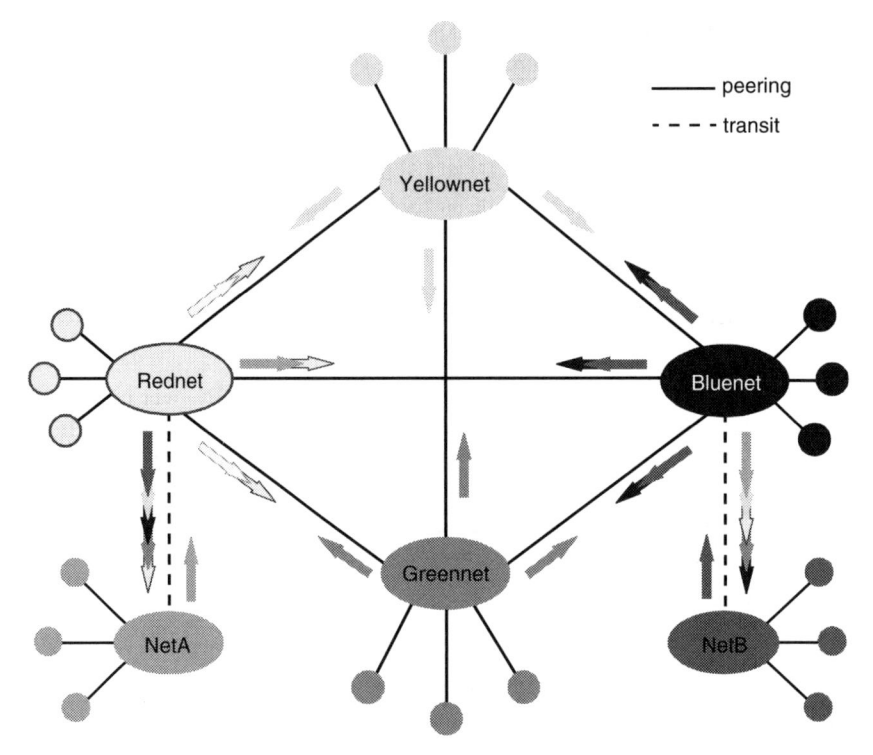

**Fig 4.5**  Rednet and Bluenet as transit providers.

### 4.4.1   Private Peering

In this scenario, shown in Fig 4.6, an interconnect, normally a cross-town circuit, is provisioned between the two network nodes. This offers a number of advantages:

- cost of UK—UK circuits are significantly cheaper than UK—USA, which will reduce the CPMS figure;
- affords good traffic control over the link;
- is appropriate where large volumes of traffic are exchanged between networks, where the cost of the circuit is less than the transit costs.

The disadvantages behind this model are:

- very slow to implement, typically months, dependent on the lead times of local carriers for the provision of the local circuit;
- it is not readily scalable, as a circuit is needed per peer;
- local carriers provide circuits at data rates that are incrementally based (2 Mbit/s-34 Mbit/s-155Mbit/s) and are priced in the same way — this often proves

*The Art of Peering* 51

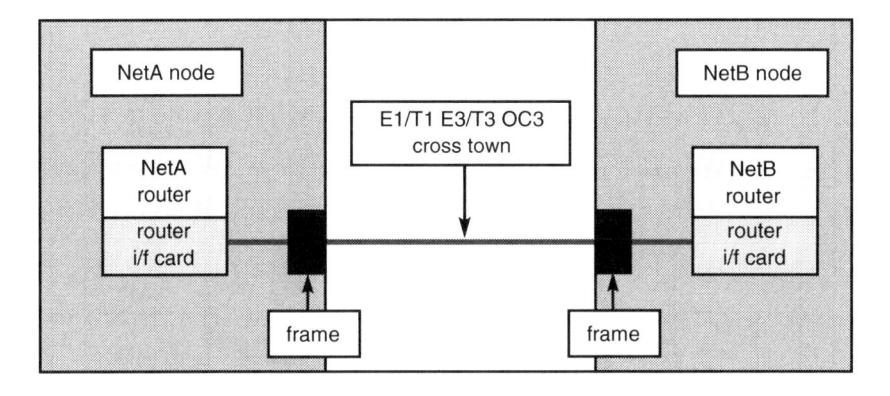

**Fig 4.6** Interconnect between two network nodes.

uneconomic on underutilised bandwidth, e.g. if the sustained traffic rates are 6 Mbit/s, a 34 Mbit/s circuit must be provisioned.

One of the drivers for private peering interconnects is the range of new QoS-based products that will not work across the existing non-QoS-enabled exchange-based structure. For instance, voice over IP (VoIP) is unlikely to gain popularity with the QoS currently available across the Internet and there is no way for any one network to unilaterally fix this. As providers strive to improve QoS within their own domains the missing piece of the jigsaw will be the interconnect space between them which includes peering policies.

### 4.4.2 Public Peering at Internet Exchange Points

Internet exchange points (IXPs) or network access points (NAPs) provide facilities which aim to keep Internet traffic within a country or region and thereby improve performance. In Europe most IXPs are operated by academic institutions or by organisations set up on behalf of ISPs in that country. In most cases these are founded on a cost-recovery non-profit-making basis for the benefit of the Internet community. This does not mean lack of organisational professionalism and support or lack of size. London, Amsterdam, Stockholm, Frankfurt all have exchange points with over 100 connected networks. Nominal revenues are derived from membership fees, which provided the exchange point infrastructure and support.

Figure 4.7 shows a typical Internet exchange point. An ISP will buy a circuit to the IXP often connecting to a router that is within their AS domain. That router is then connected to the common exchange media (typically Ethernet). Peering sessions are then set up between members over the exchange point. Peering at the exchange points adopts one of the following principles:

- bilateral peering — where ISPs can select with whom they wish to peer;

## 52  Peer with Whom and How

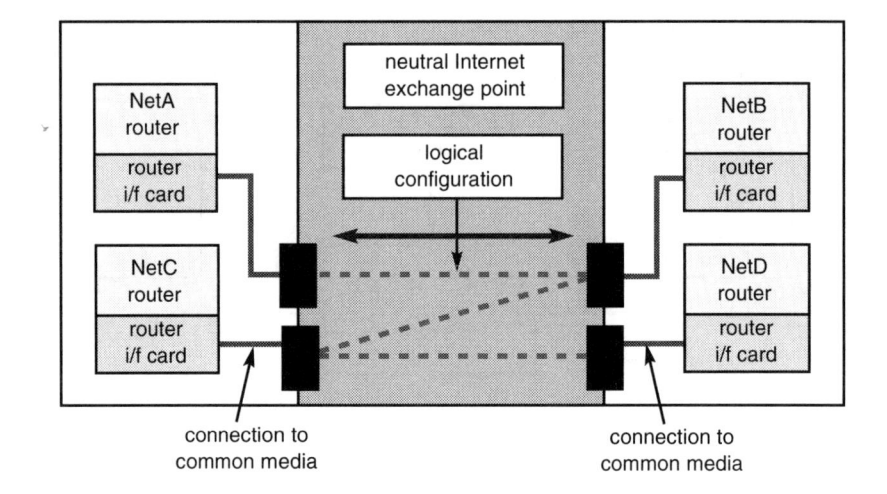

**Fig 4.7**  A typical Internet exchange point.

- multilateral peering — where all the ISPs connected must peer with all the other connected members (this latter model is now rapidly disappearing).

Public peering at exchange points provides good aggregation of traffic, and is appropriate and very cost effective where there are lower volumes of traffic exchanged between many peers. The time taken to implement peering at these exchange points is very fast, as it only requires a change to the router configuration and is not dependent on lead times for circuits.

Concert uses a combination of private peerings and bilateral peerings at exchange points throughout Europe. The BT UK service uses a combination of private peerings and bilateral peerings at the UK exchange points in London, Manchester and Edinburgh. The peerings at the LINX, the exchange point in London, are not limited just to UK-based peerings but also include those between pan-European networks such as Carrier1, Telia, Level3 and Concert.

## 4.5  Peer with Whom and How

So it can be seen that interconnection between networks must happen for the Internet to work and that the commercial mix of transit and peering improves performance and can reduce cost. The decision now is with whom and how. This leads to a peering strategy and will be based on the ISP's present market position, where it would aspire to be, and what services it is offering customers. There are three very broad segments in the market today within which ISPs are operating. These are categorised in the following sections.

### 4.5.1 Transit ISPs

Operators in this segment are operating very high-speed backbone networks within or between geographic regions. These carriers principally sell 'transit' to smaller ISPs, Internet content providers (ICPs) and very large multi-national companies. Concert's CIP service is a typical transit ISP. This segment includes the tier-1 carriers discussed earlier. Operators in this segment would peer with other operators in the segment but would like operators in the other segments as customers. Most if not all have policies which require potential peers at a minimum to have similar size networks, customer bases, and geographic coverage. There are often minimum requirements on the size of the interconnections (for example, OC3 (155 Mbit/s) and greater) minimum requirements for the number of interconnects (for example, in at least four locations), requirements on the traffic balance (for example, carrying equal traffic volumes to their respective customer base in both directions). These would be based on private peering models outside the exchange points, since exchange points may become, or be seen as, bottle-necks for the large amounts of traffic exchanged between transit ISPs. Examples of Concert peerings in Europe are Ebone, UUnet and Carrier1.

### 4.5.2 Access ISPs

Access ISPs typically operate in one country offering dial and lease line access to end users and transit services to smaller ISPs, corporates and content providers in the same country. As readily seen in the industry, operating in one country does not necessarily mean small — there are many examples of regional access networks (e.g. BTnet) with very high capacity networks, thousands of fixed connections and hundreds of thousands of dial users. ISPs in this segment would typically peer with most of the other access providers operating in the same country at the exchange points. They will switch to private peering arrangements when volumes of traffic dictate. BTnet is a typical access network with peerings at exchange points with the majority of other UK ISPs and private peerings with the larger UK ISPs. As the Internet is USA centric, any network outside the USA is an access network to the USA, as are non-tier-1 networks in the USA. The only true transit ISPs are the tier-1s; however, there are many pan-regional networks, for example Ebone and Concert in Europe, that offer transit services. The pan-regional networks offer important pan-regional connectivity as well as connectivity to the tier-1s in the USA.

### 4.5.3 Applications ISPs

This segment encompasses providers who offer Web-hosting facilities, server farms, network-centric applications, such as messaging and eCommerce, and collaboration applications. They will attempt to get interconnection with as many networks as possible to bring the content services close to end users. The traffic levels leaving these networks will make peering with these providers attractive to operators in the other two segments.

## 4.6 Peering Policy and Process

The details of the peering strategy will be driven by many factors, not just cost; these other factors may include performance, market positioning, the size of the ISP, the number and types of its customers, alliances and partnerships, and its corporate strategy. The peering strategy is a key element in determining the competitiveness of an ISP. In support of the peering strategy, the ISP will need to have a policy and process which applies consistency to a request to peer from another ISP.

As an example, it may outline for a transit provider a set of requirements such as:

- interconnection with the main in-region backbones in at least four locations, in separate countries (or states);
- peering may be implemented either privately or at the Internet exchange points;
- interconnection should be at OC3/OC12 bandwidths at least;
- balance of traffic.

As an example, for an access provider with European region peering principles the requirements would be:

- to peer with other Internet service providers within the UK for the benefit of their customer base;
- to aim to peer with networks of similar size, customer base, similar network deployment and where traffic is exchanged between the networks;
- in some circumstances (Web-hosting networks) they may seek to peer in one location based on traffic volumes;
- their preference is for public (bilateral) peering at the main exchange points;
- public peering will be based on a no-settlement model for traffic shipped.

## 4.7 Conclusions

This chapter has provided an overview of peering and the decision-making process and strategy that has an impact upon Concert's and BT's Internet services. Concert and BT seek peering arrangements to reduce transit costs and improve performance, but does not necessarily seek to peer with all networks at the expense of potential revenue. Selection of location is based on cost-effective utilisation of expensive resources. Clearly any strategy will need to accommodate changes in market conditions and growth.

# 5

# HOW TO PHYSICALLY BUILD AND MAINTAIN PoPs AROUND THE WORLD

## T Brain

## 5.1 Introduction

This chapter summarises the key aspects of building a point of presence (PoP). It presents what must be considered when selecting and placing equipment in a site and describes the options available. This chapter could be thought of as describing the layer 0 of a network, the basic foundation upon which the rest of the network and all IP services are built.

No one can plan to build a carrier-scale IP network without first considering accommodation, air conditioning, power, access to bandwidth, equipment delivery and installation, all of which are summarised here.

An essential part of being a global carrier-scale IP network provider is the ability to install and maintain PoPs right across the world. There are two ways of building an international IP network — either by buying and absorbing existing regional networks, or secondly, by building your own. The synergies between an in-country IP network and a global IP network are readily apparent — both rely on the same fundamental hardware and both are typically based on broadly the same hierarchical infrastructure design. However, once you expand outside the confines of your native country those synergies start to dissolve, leaving you with a host of country-specific problems, from equipment being held up in customs, to a lack of suitable power supplies or bandwidth.

## 5.2 Location

As with all disciplines associated with building global networks, planning is no less difficult than any other and will, if not correctly carried out, add to the cost of the

network. When planning a network one of the important factors that must be explored from the outset is the location of the network PoPs. This area of planning must be the most singularly important aspect of the build, and, as such, needs careful consideration. An incorrectly chosen site will have a knock-on effect (mostly in terms of additional engineering and operational costs) for both the existing network, and for any expansion and growth later on in the network programme.
Site selection generally comes under three headings.

- Collocation

    This is where a site is shared and typically controlled by an in-country partner or distributor. Over the last few years deregulation has opened up the global IP markets for expansion. However, this does not automatically result in an operating licence being issued to just anyone. One way around this is to go into partnership with an existing in-country Internet service provider (ISP) or licensed operator. In the cases where this happens, it is usually more convenient to enter into some form of collocation agreement with that operator.

- Telehousing

    This is where space is rented within a purpose-built building, shared with other service providers. The advantage of this is that those other service providers can also become your customers or suppliers. Being in the same building also reduces the need for expensive interconnecting bandwidth. Telehousing does have the possible compromise of security, and visibility to competitors of the equipment used.

- Purpose-built

    This is the last option and is where a site is purpose-built to support a specific product or products. Global communication companies will use one or more of the above options depending on the size of their operation. IBM, prior to the sale of their IP network, predominantly collocated their equipment within their in-country office facilities. Concert, however, uses a combination of all three types of accommodation, as part of its global expansion.

It is generally accepted that the purpose-built option gives greater long-term opportunities, not least of which is the ability to offer telehousing facilities to other providers or customers.

The location of the PoP must be such that it is within easy reach of the country's and/or city's Internet exchange points (IXPs), and obviously near to where major large customers are likely to be sited. The building itself must also have a long-term future and include all the environmental facilities that a growing network will require, including ready access to suitable bandwidth, and local/international circuits that are capable of supporting projected growth.

## 5.3 The Environment

To get the optimal performance out of the equipment, careful consideration is needed regarding the environment in which the equipment will be housed.

A typical equipment room will have dual cable entries for diversely routed cables (for international and local connections) into the building. The floor area should have a 'clear height' for rack areas, of not less than 2.85 m. Access into the room should be available either via a loading bay or goods lift. The minimum opening for doors and lifts is 1.6 m wide × 2.45 m high.

Installed equipment requires an air-conditioning system capable of maintaining a temperature of 24°C ± 3°C (75°F ± 4°F ), with a relative humidity of 50% ±10%. Even with a system that maintains the ambient surroundings at the correct temperature, hot spots may still occur within the equipment racks themselves if airflow through the rack has been restricted in some way.

Modern routers have high-level temperature alarms that will shut down the router if the processors get too hot (see Fig 5.1).

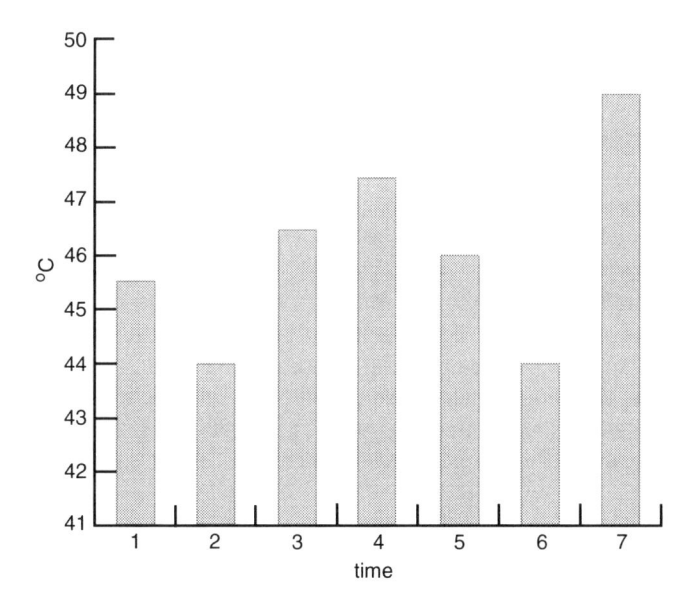

**Fig 5.1**   Example of CPU temperature rise in a badly ventilated cabinet.

## 5.4 Power Options

Power options fall into two areas — either AC or DC. The AC option will be country specific. Within Europe 220 V is typical but in Japan 100 V is the standard. This can give problems when deploying equipment such as the Cisco 12012 router whose power units do not auto-range down to that level. Where this is an issue, special arrangements have to be made to connect two mains phases together to increase the output voltage. When doing this, great care must be taken to ensure that other equipment within the rack area is not jeopardised.

The DC option originally derives from the telephone exchange practice of having a central battery. For smaller installations these were replaced with power equipment racks (PERs). These racks contain a number of lead acid batteries and a floating charge arrangement to give a continuous output of $-48$ V DC.

The simplest and most common solution is the AC option supported by an uninterrupted power supply (UPS). The UPS is an on-line solid-state item of equipment which operates both from the public AC mains supply and from a stand-by generator.

It consists of a rectifier, emergency stand-by battery, a solid-state inverter, a solid-state continuously rated static bypass switch, an internal maintenance bypass switch, and a fully automatic control and synchronising system.

The function of the UPS is to ensure a continuous supply of a quality alternating power source. The UPS normally has a mean time between failure (MTBF) rate of at least six years.

## 5.5   Protective and Silent Earths

To prevent cross interference from other equipment both protective and silent/signal earths are required. The silent earth should be bonded back to the main building earth using 70 sq mm cable.

In the equipment room both the raised floor system and overhead cable tray system, if used, will require bonding to the protective earth. Parallel paths are provided to ensure a low impedance path at high frequencies.

Each cabinet will require independent bonding to the cable tray system. Bonding conductors are sized for high frequency (1000 Hz to 2000 kHz) grounding. Braided or flat copper straps are the preferred conductor for high frequency bonding and ground jumpers.

## 5.6   Earthquake Bracing

In areas prone to earthquakes, such as the western seaboard of the United States, or Japan, additional bracing is required to prevent damage to the equipment racks. This can take the form of a specially designed plinth which is fitted below the rack within the raised floor cavity.

The extent of this bracing will be dictated by local practices. These same practices may also dictate the maximum amount of power that any one building or equipment area, may safely deploy.

## 5.7 Bandwidth Ordering

Bandwidth is the single most expensive part of any global network and costs many millions of pounds per year. Typical international private circuits (IPCs) within Europe cost in the order of $20 per kbit per month, while in the Asia/Pacific area $29 per kbit per month is the norm. Regional initiatives, through deregulation, have greatly reduced these costs within Europe, and many of the larger players, such as Cable and Wireless, are now actively laying their own fibre systems for long-term investment. BT along with its partners already operates a European backbone network called Farland (see Fig 5.2). This allows its members access to pan-European bandwidth at greatly reduced costs. Another consortium, Hermes, utilises fibre optic cables laid alongside railway lines. Although great inroads are taking place within Europe, the Asia/Pacific region is still lagging behind, but by the close of 2001 similar cable networks are expected to be completed that will open up the global expansion of IP networks on a vast scale.

## 5.8 Equipment Ordering

A good working relationship between the network provider and their equipment supplier is vital to ensure the timely delivery of network components. It is equally vital that all equipment destined for a site is homologated for use within that country. Furthermore there are a number of countries that are export blacklisted by the US Government. This prohibits the export of certain high technology equipment to certain parts of the world. These restrictions include the latest router equipment from suppliers such as Cisco.

The control of logistics and material management is an area that should be conducted through a single point of contact such as a project manager. This will ensure that the deliveries of set milestones are complete prior to moving on to the next critical step.

Some equipment manufacturers will only deliver within the country of manufacture or alternatively the product manager may wish to control the shipping himself. In these cases a freight forwarder may be used.

The freight forwarder will handle and complete documentation world-wide on behalf of the network provider. This covers not only documentation and information as required by the relevant international customs authorities, but also any licensing regulations and conditions, payment of appropriate duties and taxes, including VAT, when requested and allowed by the importing country. The freight forwarder will also claim any allowable airfreight apportionment discounts on the shipment transactions. The freight forwarder also takes care of any problems relating to declarations made on behalf of the network provider to customs. Within the United States of America, freight forwarders are given a valid power of attorney to allow them to make tax and duty payments on behalf of the network provider.

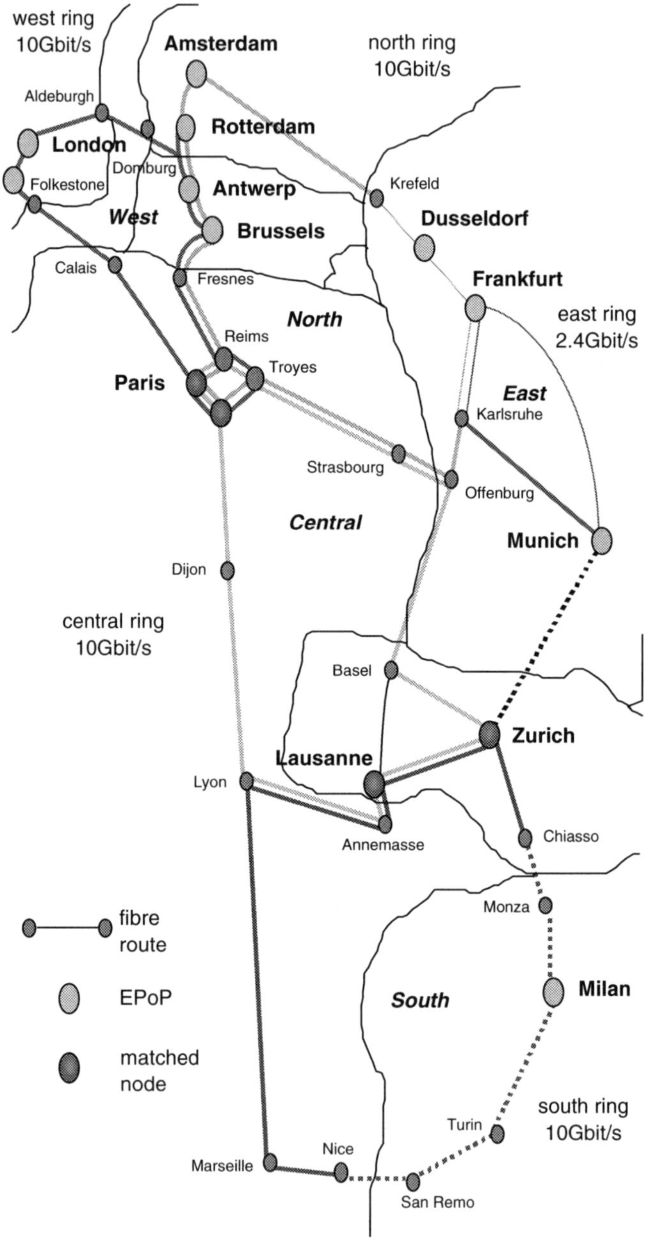

60  *Equipment Ordering*

**Fig 5.2**  The 'Farland' network.

## 5.9 Equipment Delivery

There are two approaches to installing equipment on a global scale. The first is to have all components delivered to site, either directly from the equipment supplier or via a freight forwarder.

The second is to have all equipment sent to a single location, checked out, built as best as possible, then shipped off to its final destination via a reputable courier. Each approach has its own merits.

When building a global network, getting equipment on site can be a logistical challenge. Notification from suppliers commonly does not match delivery notes of what has actually turned up on site. So firstly delivery of kit needs to be confirmed by someone who knows what it is they are checking. Once delivery has been confirmed a team of engineers may be despatched. If there are any problems with the kit during the installation, an agreement with suppliers should be reached to ensure that any equipment that is delivered 'dead on arrival' (DOA) is replaced within 24 hours. This will allow the full installation to be completed while the engineers are present and save a return visit.

Racking and stacking can help resolve this problem by having the equipment sent to a common location where delivery checks are confirmed. The equipment is unpacked, tested, and installed in racks prior to onward shipping. This method ensures that all equipment sent to site is correct and working, although damage through transit is not unknown. However, this method also adds extra costs to the installation process, and where tight deployment schedules are an issue, can introduce unacceptable delays.

## 5.10 Equipment Installation

Building the node can be an impossible task if qualified installation engineers are unavailable. Hiring temporary local staff is one solution, but is not always an option and can be expensive. Installations can also be carried out by using the network provider's own staff, or by using a third party 'turn-key' contract. There are clear advantages and disadvantages for each choice.

Having trained experienced staff would provide the added knowledge that all installations are being carried out to a known standard. These staff can also be called upon for ongoing changes and maintenance; however, if just used for installation, this would be a large overhead unless utilised to meet the day-to-day operational requirements as well. Also, using local staff ensures that local difficulties are resolved much more quickly, and travel times for staff are kept to a minimum. Turn-key staff on the other hand alleviate the need for full-time, experienced staff to be employed other than for the period of the installation.

## 5.11 Sparing

As part of the initial equipment purchase it is normal for the network provider to enter into some form of repair and maintenance agreement (RMA) with the supplier. This will enable faulty equipment to be replaced in a timely manner. Because it is vital that network outages are kept to a minimum, an on-site spares holding is required for customer-service-affecting equipment. To supplement on-site holdings, global IP networks require operational, and non-customer-service-affecting, spare equipment to be available to its network locations within 48 (exceptionally 72) hours. Stocks can be held at regional depots from where they can be deployed to site as required. As with most electronic cards, the equipment is sensitive to static discharges and must be handled carefully to prevent damage. The regional centres can arrange appropriate packing for shipments. They also ensure that the equipment movements comply with the appropriate regional and in-country import/export regulations.

Normally three regional depots, serving three regions, namely Europe, Americas, Asia/Pacific, will suffice, although the location of these depots will be influenced by the practicalities of achieving the defined target delivery time-scales.

The regional centres will also track the RMA process, ensuring timely repair of faulty equipment and delivery of replacements to site.

## 5.12 Operational Access

There are two ways of operationally accessing an IP network — in-band, using the existing core infrastructure, and out-of-band (OOB). For in-band access an operations centre will need direct connection to one or more of the network nodes. This allows immediate indication of alarms within the network and is the most common method of network monitoring. Some, but not all, networks supplement this arrangement with OOB access either via a PSTN or via an ISDN connection. For most applications the PSTN access via a modem is more than sufficient to cater for any additional access to the nodes. ISDN on the other hand can provide a cleaner line for data interchange. The main disadvantage with ISDN, apart from costing more than a PSTN connection, is that a supplier normally delivers an in-country variant as default as opposed to an international standard. Also, as there are few terminal adapters that have full regulatory approval in all countries, typically two types will be required for a global network. A third OOB option exists for the larger networks in the form of a global management LAN. This is an overlay network that allows not only OOB access to IP networks, but also management access to any other telecommunications products offered by that network provider. It is important to remember that the prime function of any OOB system is to provide access during a network outage, and is, as a consequence, used infrequently. Therefore a routine testing programme must be implemented to ensure that all OOB access is fully functional at all times.

## 5.13 Operational Cover

The best form of operational cover is given by on-site staff operating on a 7 days by 24 hour rota. An added benefit to this arrangement is that, where such a staffing arrangement is applied, this resource can also be used for the installation and day-to-day changes to the equipment. The overheads for a medium-sized site are more than covered by the savings in travel and accommodation costs that would otherwise have been incurred by separate installation and service delivery teams making specific journeys to the nodes. This is more so for organisations that offer multiple products such as frame relay, ATM, IP and voice services.

For overall operational control a single operational centre is all that is required for a regional network. However, for a truly global network this centre needs to be supplemented with at least one additional unit. Ideally a 'follow-the-sun' policy should be adopted with one centre having overall control, but handing the day-to-day functions over to the other centres based on local time of the day.

Operational centres are not only responsible for alarm monitoring but also have responsibility for the logical builds and security of the network.

## 5.14 Conclusions

This chapter has described the principles of physically building global networks and covered the practical details learnt by Concert. A key contributor to Concert's success is its ability to plan and provide points of presence around the globe. The challenges and details highlighted in this chapter should not be overlooked when planning a global network.

# 6

# UK CORE TRANSMISSION NETWORK FOR THE NEW MILLENNIUM

## I Hawker, G Hill and I Taylor

## 6.1   Introduction

Those building IP networks see two types of connection as key to them — the access connections, the cables, fibres or wireless systems that connect their customers to the PoP, and the core connections, the fibres that connect the PoPs together. This chapter describes the latter aspect, looking at today's and tomorrow's transmission technology. Utilising cost effectively the capacity of the fibres buried in the ground and being able to reconfigure the transmission network in a timely manner to meet rapidly growing customer demand are of interest chiefly to ISPs and carriers that have their own transmission network. However, the cost of installing and managing the transmission network is a fundamental cost that is inherited by the IP networks.

The pace of change in telecommunications networks, driven by information technology, continues to accelerate, resulting in a whole new range of narrowband and broadband data services. This has placed new requirements on the underlying transport infrastructure (SDH, WDM, and fibre) in terms of capacity growth [1] and management, where BT spends some £1bn per annum on new equipment, operations and support. Whereas new operators purchase the latest technology 'off-the-shelf', rolling out 'green-field' networks, BT and the other established operators face the challenge of interworking with their existing network.

Using the development of the BT core transport network as a case in point, the chapter discusses long-haul and metropolitan networks, but excluding the last mile from the local exchange to the customer. An overview of current network developments in SDH and WDM is provided, and future network opportunities, driven by advances in technology and the continuing need to reduce costs and improve performance, are also discussed.

## 6.2 Drivers

A number of key drivers are shaping the BT/UK core transport network (see Fig 6.1):

- higher bandwidth access networks, using ADSL (asymmetric digital subscriber line) technology for small businesses and access SDH for large businesses, are being deployed widely to provide broadband services such as data transfer and video to several million users within a few years — this will increase core traffic at least tenfold;

- large increase in international traffic from new sub-sea cables, such as TAT14, which require 300–600 Gbit/s traffic to be backhauled into London from frontier stations in Cornwall by September 2000;

- new capacity to support leased lines for other licensed operators (OLOs), reflecting high growth in the wholesale market;

- new 'Cyber Centres' within metropolitan areas (especially London) housing very large IP routers for eCommerce, where traffic will be especially unpredictable as new services are created and deployed quickly by Internet service providers (ISPs);

⇒ market volatile, competition high
existing network becomes stressed . . .

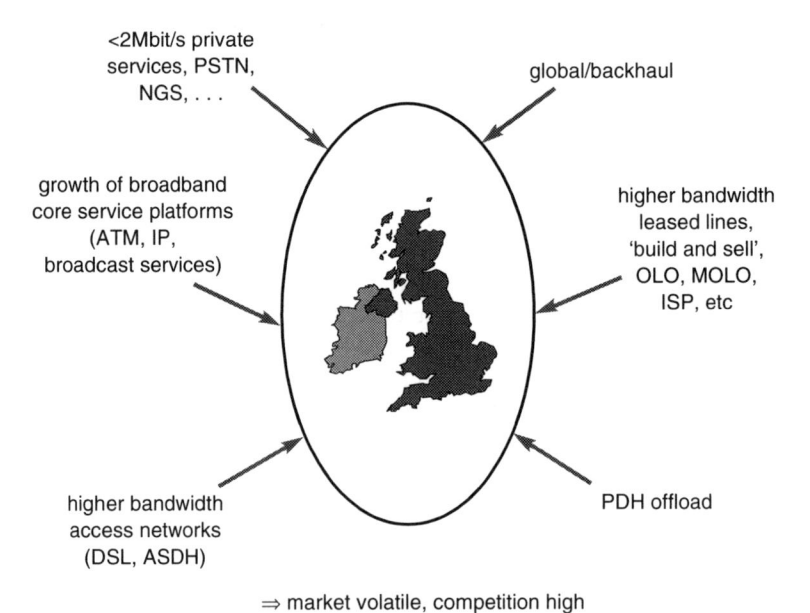

**Fig 6.1** Drivers for network change.

- overall an average traffic increase of about 100% per year can be expected (Fig 6.2) plus large geographic variations;
- unit transport costs must decrease inversely with traffic growth for a given level of investment — this is being achieved by introducing new technologies such as 10 Gbit/s SDH SPRings and 40 Gbit/s WDM point-to-point systems (16 × 2.5 Gbit/s), with the promise of 40 Gbit/s SDH and 400 Gbit/s WDM systems (40 × 10 Gbit/s) in 2001; in the future, simplified platform architectures, running, for example, IP over WDM, will further reduce unit costs;
- greater flexibility and scalability is needed at all bit rates to allow the network to efficiently transport, on demand, traffic of increasing unpredictability (e.g. Cyber Centres and start-ups) — scalability is essential in terms of circuit numbers and management where automated systems may grow to any size over the next five years;
- network reliability must continue to improve, providing to customers the option of hitless services following network failures such as cable breaks — this implies a high degree of automation and preplanning within the OSS and managed pool of spare capacity;
- bandwidth transparency — emerging data platforms generate new requirements to support large 155 Mbit/s+ chunks of capacity granularity from IP routers and ATM switches, making WDM attractive since it is largely bandwidth independent — over time, SDH will become a single client among others (IP, ATM, etc) running over a terabit WDM infrastructure;
- the network must have managed flexibility for all granularities to allow new circuits to be simply set up either by the operator or by the customer (for

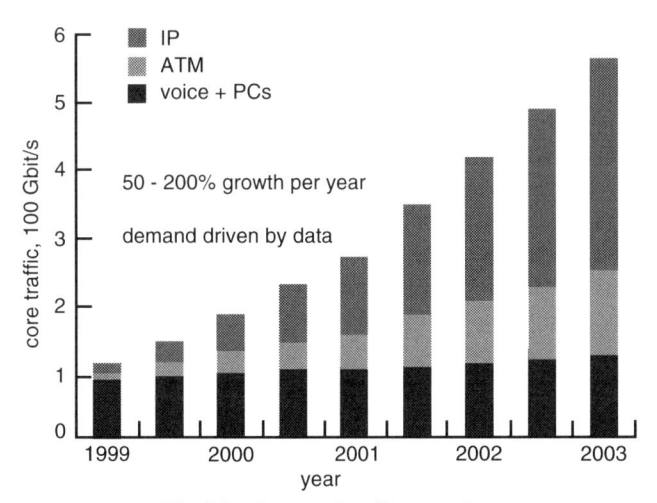

**Fig 6.2** Expected traffic growth.

instance, for customer private networks), e.g. it may be necessary to divert circuits temporarily (days or weeks) for upgrades or because cables need to be re-routed around roadworks — also, it should be possible periodically to re-optimise the network routings without affecting customers, because increases in volume and changes in demand patterns make new routes more economic (the management systems must help to identify such opportunities for optimisation and to automate the rearrangement process as far as possible);

- there is a growing demand from large customers for leased wavelengths within both the UK and Europe, which will initially be provided over WDM point-to-point systems between major nodes — later additional flexibility can be achieved by optical networking, allowing wavelengths to be assigned, on demand, across the network from an operations centre.

In general advanced network management features such as 'auto-discovery' (automatic recognition of a new element or feature by the network itself) will help contain management costs as networks grow. Also a holistic approach is needed to network development, spanning all of the network and client layers so that it becomes possible to 'mix-and-match' platforms as required (for example, run IP directly over WDM if applicable) to further reduce unit costs. This approach also implies better interworking of management systems across platform domains.

## 6.3 Current Transport Network

In the last ten years SDH technology has been rolled out to more than a thousand nodes, enabling development of a software managed transport layer which can be configured, monitored and optimised from a network operations unit (NOU). This technology contains building blocks such as add-drop multiplexers (ADMs), cross-connects (XCs) and line systems, bolted together in various configurations to meet the requirements of a particular sub-network. Recently wavelength division multiplexing (WDM) technology has been introduced to increase capacity using many (16+) wavelengths per fibre (rather than a single wavelength) to reduce link costs.

While SDH has been built over the last ten years, BT has still increased significantly the size of the PDH network and has only recently closed the FDM network. This chapter, however, focuses on the new SDH/WDM technology.

The current SDH transport network (Fig 6.3a) has developed primarily to support 2 Mbit/s private circuits and more recently the emerging broadband platforms. It has several layers:

- a long-haul network (tier 0) to backhaul sub-sea traffic from the frontier stations to international switching centres in London — generally high-capacity 40 Gbit/s+ SDH/WDM point-to-point systems are used, hubbing to the major international gateways at London and Madley;

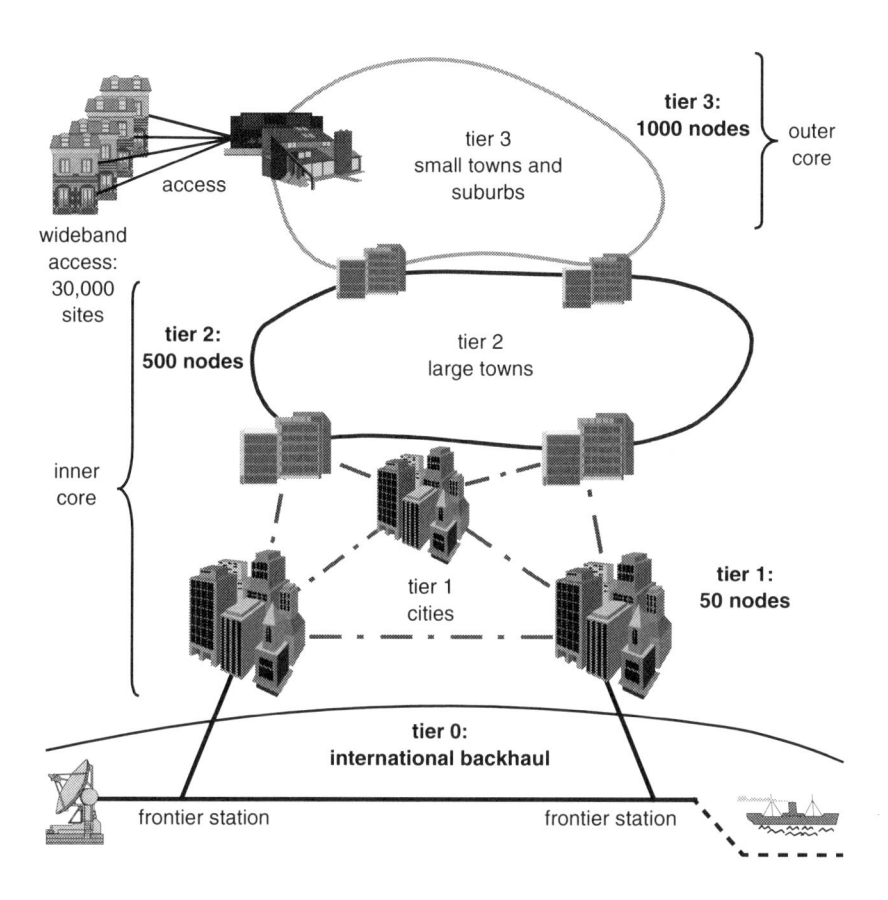

*UK Core Transmission Network for the New Millennium* 69

**Fig 6.3a** BT/UK network hierarchy.

- an inter-city meshed network (tier 1) connecting about 60 major nodes (Fig 6.3b), for which a high degree of networking and connectivity is needed in order to allocate capacity on demand and to provide a portfolio of network protection options — in addition, an overlay of SDH SPRings is being deployed (Fig 6.3c) to support broadband data traffic generated by the developing MSP and DSL platforms (known as tier A/B rings respectively);
- super-cells between tiers 1 and 2, providing interconnection of several tier-2 rings, were needed, especially at the network periphery;
- regional networks in tier 2 connect approximately 500 towns on to SDH rings;
- rural and city SDH rings in tier 3 are currently being deployed, connecting about 1000 local exchanges — these are largely infill within city and urban areas and extend the reach of SDH into the outer core;

70  *Current Transport Network*

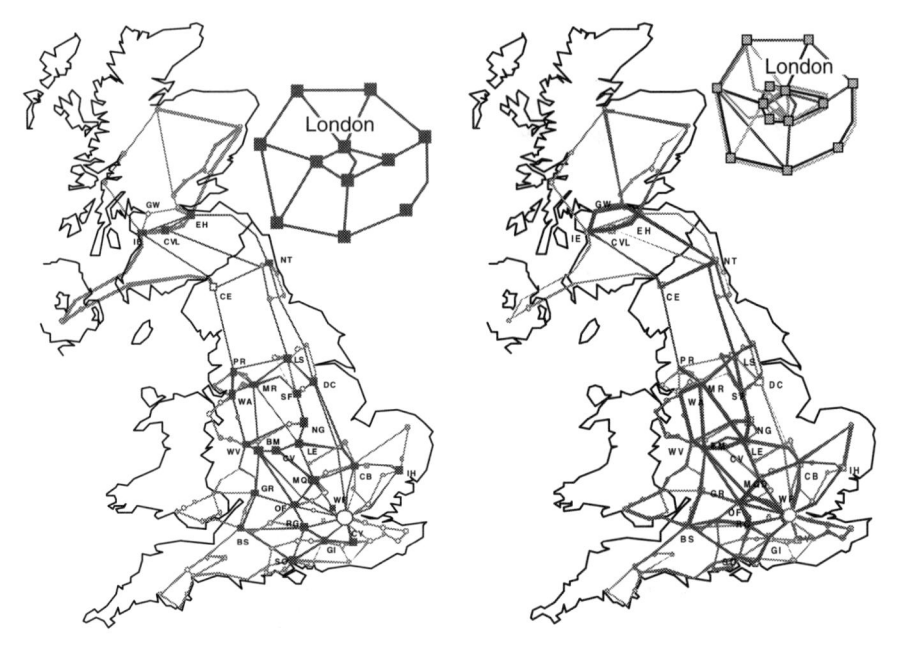

**Fig 6.3b**  SDH meshed network (tier 1 and supercells only).

**Fig 6.3c**  SDH SPRing overlay.

- SDH direct to more than 1000 large customer sites at 155 Mbit/s and above (retail and wholesale);
- connection to 200—500 OLO sites (national and international).

The existing network architecture (see Fig 6.4) for the core network has two major components:

- a well-established narrowband VC12/VC4 mesh layer using large XCs/ADMs supporting 2/140 Mbit/s private services and related products;
- a new broadband VC4 layer to support mainly IP/ATM data traffic with built-in 50 ms protection using SDH SPRings.

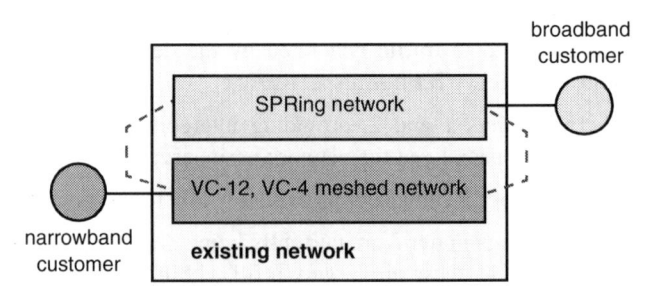

**Fig 6.4**  Existing network architecture.

The management systems for these layers are being connected and automated allowing circuits to be routed seamlessly across the core network

Table 6.1 illustrates how SDH technology is developing in terms of the capacity of SPRings and XCs to support 2 Mbit/s and 140 Mbit/s private services and data.

These developments will continue to reduce unit costs and increase network capacity over the next few years.

**Table 6.1** SDH development.

| Date | SPRing capacity | Cross-connect capacity (2 Mbit/s) | Cross-connect capacity (140 Mbit/s) |
| --- | --- | --- | --- |
| 2000 | 10 Gibt/s | 80 Gbit/s | 300 Gbit/s |
| 2001 | 40 Gbit/s | 250 Gbit/s | 450 Gbit/s |
| 2002 | 40 Gbit/s+ | 250 Gbit/s | 600 Gbit/s |
| 2003 | 40 Gbit/s+ | 250 Gbit/s | 900 Gbit/s |

In addition BT has 20–30 WDM point-to-point systems deployed on major routes such as London—Birmingham:

- to enhance the capacity of existing cable systems by increasing the number of wavelengths from 1 to 16 per fibre pair;
- to directly support the IP core backbone (Colossus) without using SDH switches (interfaces remain SDH).

Currently each WDM system has up to 40 Gbit/s of capacity with 16 protected wavelengths at 2.5 Gbit/s per wavelength and future WDM systems will be expected to provide at least 40 wavelengths at 10 Gbit/s per wavelength. Use of lower cost WDM for metropolitan systems is also under investigation.

## 6.4 New Network Developments

Having examined the major drivers for the next few years, some likely network developments are now considered, including deployment of a new broadband layer for transporting the next generation of broadband and Internet services.

### 6.4.1 A New Broadband Layer

The current SDH network has been designed to support 2/140 Mbit/s services. For new IP and ATM data services, a new broadband overlay network will be deployed from April 2001 onwards (see Fig 6.5) to provide a much lower cost transport for data at 155 Mbit/s + granularity. Existing 2-Mbit/s traffic will remain in the existing narrowband SDH network. Eventually this new broadband layer will become 'all-

72  *New Network Developments*

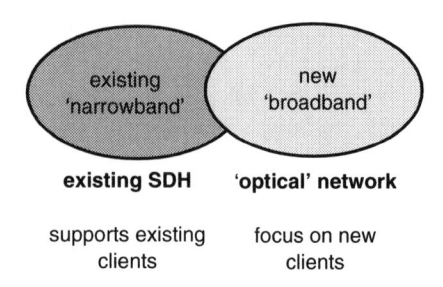

**Fig 6.5** A new broadband layer.

optical', providing wholesale wavelengths to support individual client IP, ATM or SDH client networks. The new transport network will also make use of more flexible platform architectures and the most recent very high capacity transmission nodes.

### 6.4.2  New Platform Architectures

Currently, almost all services and platforms are supported over SDH (or legacy PDH) including private circuits and data. However, while SDH provides QoS guarantees, this is not always necessary for IP data services such as e-mail and World Wide Web. There is also a trend for routers to offer quality of service (QoS) ranging from best effort e-mail and Web searches through to guaranteed electronic commerce.

The emerging architecture (Fig 6.6) is flexible allowing platforms to be 'mixed and matched' according to the client layers being supported. On day one, platforms tend to be stacked vertically on SDH as the common bearer across a wide service mix. However, in time the various platforms can be arranged to support voice, broadband data and IP services more cost effectively, and a migration to simpler platform architectures can be seen, for example:

- voice services begin to migrate from narrowband PDH/SDH to an IP platform (VoIP);
- data migrates from SDH/ATM platforms to an IP/MPLS platform (discussed later);
- IP can be supported over various platforms to suit the existing infrastructure;
- direct connection of clients to the managed optical layer (via transponders) for data switches with course granularity interfaces at 155 Mbit/s +.

The major challenge as always is to develop and interwork the network management systems (discussed later in section 6.9).

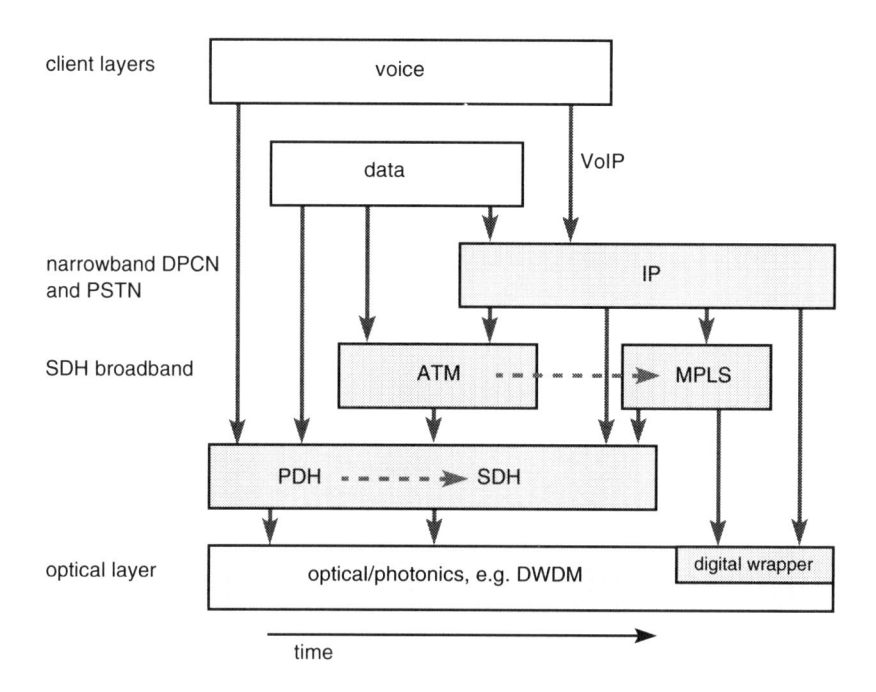

**Fig 6.6** Transport network architecture evolution.

### 6.4.3 New Node Architecture

The new broadband network will have switching nodes containing broadband SDH and optical switches, IP routers and ATM switches. Figure 6.7 indicates the architecture of a gateway node linking the long-haul and metropolitan networks. The node architectures must be scalable given the growth and uncertainty in traffic demands in the range 2 Mbit/s to 10 Gbit/s. It must also be resilient to equipment failures, fires, etc, through duplication and other means.

### 6.4.4 Layer Integration

While there are some opportunities for layer integration across IP, ATM and SDH (to reduce costs), this needs to be balanced against loss of flexibility in a fast-changing environment.

For example, IP routers are evolving more rapidly than SDH switches in terms of size, functionality and cost. Currently there are possibilities for layer 3/layer 2 integration using MPLS, especially at the core network periphery.

74  *Reducing Unit Costs*

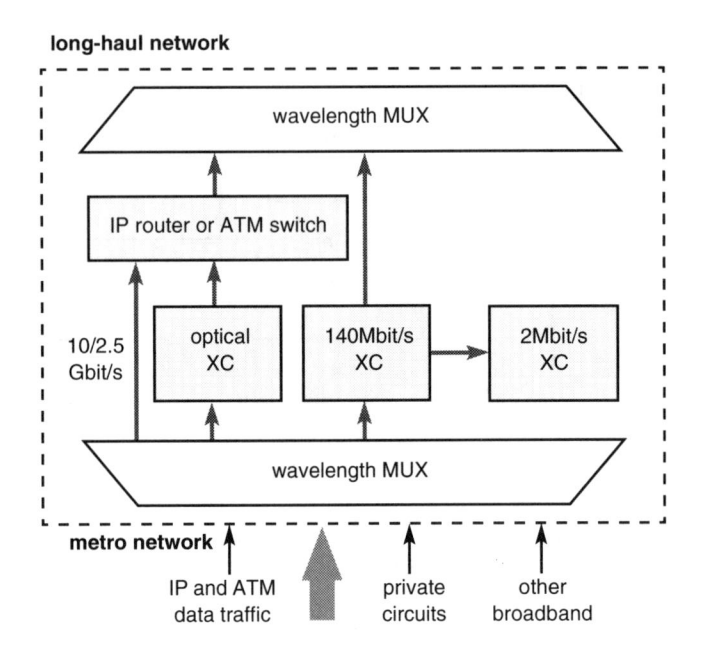

**Fig. 6.7**  Broadband node architecture.

## 6.5   Reducing Unit Costs

The architectural simplifications described above, together with higher bit transport rates, will reduce unit capital costs significantly as the technologies are deployed over the next few years. The stages, shown in Fig 6.8, are:

- existing SDH mesh (A) core network;

- WDM (B) point-to-point systems reduce unit costs by increasing the capacity of existing cables and by replacing 3R regenerators with optical amplifiers;

- 10 Gbit/s SPRing (C) overlay reduces costs compared with the mesh by using high volume ADMs rather than large expensive XCs — interconnect of SPRings to form a network is via smaller, cheaper XCs;

- SPRings with embedded WDM (D) reduce unit costs and offer very high capacity (currently 80 Gbit/s in BT's pan-European network, 'Farland');

- optical networks (E) offer additional flexibility over WDM point-to-point and enables greater utilisation of BT's fibre assets — they will also be able to support flexible wavelength services required by larger customers.

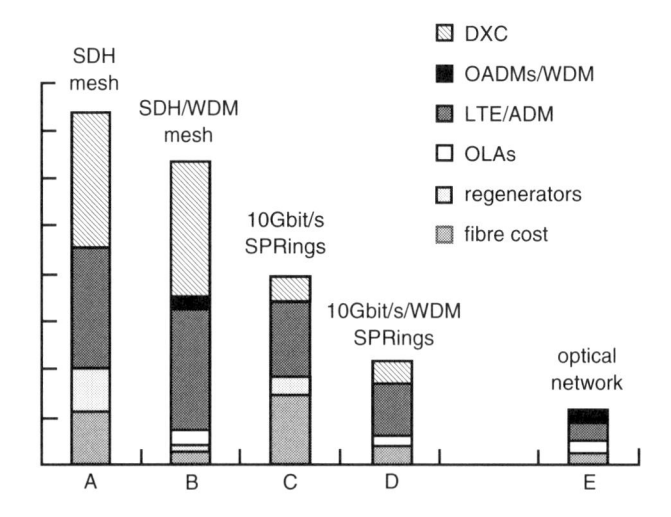

**Fig 6.8** Estimated unit cost reductions from new technologies.

The generic roadmap (Fig 6.9) based on industry trends indicates a migration route for transporting major client platforms:

- the private services platform at 2/140 Mbit/s granularity continues to be supported on SDH, which in turn evolves from 10 Gbit/s SPRings to very high capacity SDH/WDM SPRings at 80 Gbit/s+, and for which SDH continues to provide the management and QoS guarantees — in the longer term private services migrate towards other platforms, such as optical and IP, and are transported as data services (see sections 6.6 and 6.17);

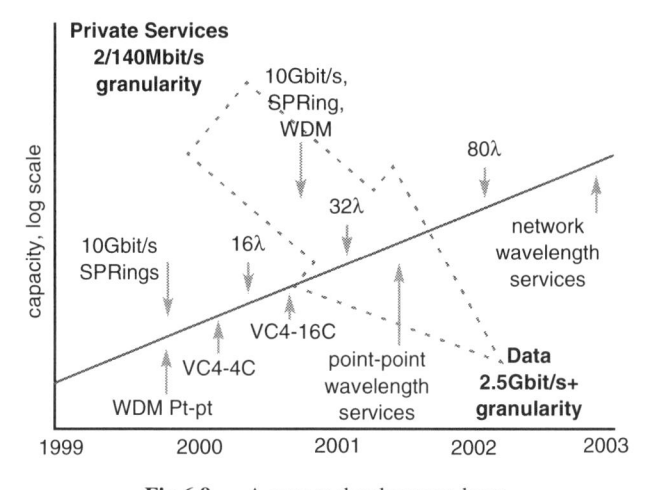

**Fig 6.9** A core technology roadmap.

76  *The Optical Layer*

- IP and ATM data services with 155 Mbit/s+ granularity are supported either on the existing SDH platform via concatenated VC4s ($n \times 140$ Mbit/s blocks of capacity) or directly on WDM systems, evolving to a managed optical platform over the next 3–5 years;
- wavelength services to larger customers are offered initially as point-to-point wavelengths between large centres, and, later, as fully networked wavelength services on demand — this will require management of the optical layer by employing developments such as digital wrappers (described later) which allow management of individual optical circuits.

## 6.6  The Optical Layer

This section considers developments in the optical layer and the migration from WDM point-to-point systems to optical networking, which is especially important within the context of new data services.

### 6.6.1  Standards

Deployment time-scales for a managed optical layer are largely dictated by standards in order to facilitate interworking and wavelength routing between operators and suppliers. Important dates for ITU are listed in Table 6.2.

**Table 6.2**   ITU standards — time-scales.

| Standard | Subject | Publication date |
|---|---|---|
| G.872 | Architecture | Published 2/99 |
| G.798 | Equipment | Mid 2001 |
| G.707 | Network Node Interface | End 2000 |
| G.959 | Physical Aspects | End 2000 |
| G.874 | Network Management | End 2001 |
| G.875 | Management Information Model | End 2001 |

The publication of a standard can mean many things — it does not, however, guarantee interworking on day one or that all issues have been dealt with on publication. Usually a standard will be published several times each time with more material.

### 6.6.2 Digital Wrappers

To effectively manage wavelengths or optical channels (Ochs) requires that optical networks support per-wavelength OAM functions. Digital wrapper technology, taking advantage of existing optoelectronic regeneration points within optical networks, may provide optical layer performance monitoring, forward error correction (FEC), and ring protection on a per-wavelength basis, independent of the input signal format.

Figure 6.10 illustrates the proposed frame structure. The optical channel OAM overhead is divided into optical section, tandem-connection and optical path, similar to SDH, and provides a similar range of functions.

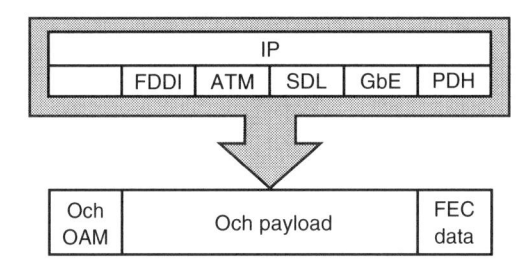

**Fig 6.10** Digital wrappers for management of the optical layer.

### 6.6.3 Optical Networking

Early optical networks in the UK will include optical add-drop multiplexers (OADMs) with both static and dynamic wavelength routing. Figure 6.11 shows a self-healing optical ring where wavelengths, supporting a range of bearers and bit rates, can be added and dropped at nodes under control of an element manager. OADMs are self-contained and represent a relatively low-risk way of introducing optical networking. Later these rings may be interconnected using optical XCs to form larger networks enabling wavelengths to be managed across large geographical areas. Some operators will be deploying optical networks aggressively in year 2001 ahead of standards (which should be about 80% in place by end 2001).

The development of optical network elements is summarised in Table 6.3 where it can be seen that optical XCs and OADMs switch increasing numbers of wavelengths. Combined with new management systems this will enable full optical networking within a few years.

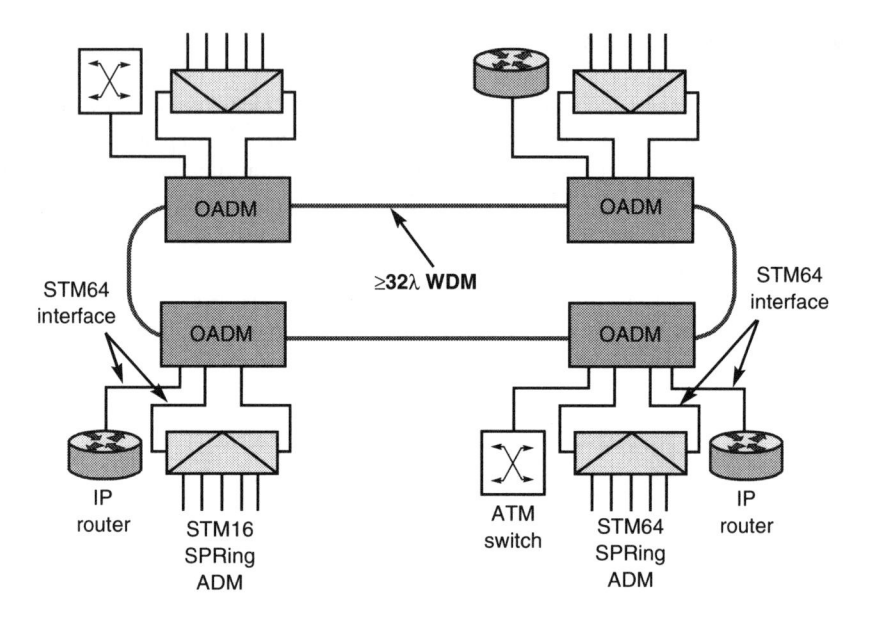

**Fig 6.11** OADM rings supporting various client platforms.

**Table 6.3** WDM developments.

| Date | Point-to-point system wavelengths @10 Gbit/s | OADM wavelengths @ 10 Gbit/s | Optical cross-connect wavelengths @ 10 Gbit/s |
|------|----------------------------------------------|------------------------------|-----------------------------------------------|
| 2000 | 80   | 40     | —    |
| 2001 | 160  | 80     | 256  |
| 2002 | 160+ | 80—160 | 1025 |
| 2003 | 160+ | 160    | 2048 |

## 6.7 Transport for IP Services

This section focuses upon transport technologies for supporting IP services where fundamental developments can be expected over the next few years.

In the US today there are approximately 50 national IP backbones and more than 5000 regional IP backbones and Internet service providers (ISPs). Some provide only traditional IP services such as dial-up for e-mail, file transfer and Web-browsing applications — which do not demand strict QoS guarantees from the network and are not sensitive to delay. Conversely, other providers offer additional services such as Web hosting and mirroring, eCommerce, and emergent services such as virtual private networks (VPNs), IP telephony and multimedia services.

These applications do require strict QoS guarantees including, for example, maximum roundtrip delay, low packet loss and scalability to a massive number of users.

### 6.7.1 Adaptation of IP to WDM

There is no single method of supporting IP over WDM and it is useful to review the various mechanisms used today and proposed for the future. Referring to Fig 6.12, methods (a), (b) and (d) are most commonly used today.

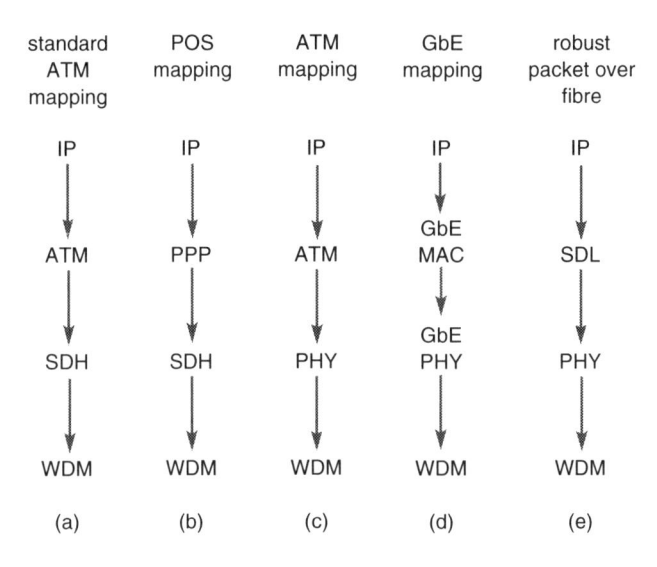

```
standard      POS         ATM         GbE        robust
 ATM        mapping     mapping     mapping    packet over
mapping                                           fibre

  IP          IP          IP          IP          IP
  ↓           ↓           ↓           ↓           ↓
                                     GbE
 ATM         PPP         ATM         MAC         SDL
  ↓           ↓           ↓           ↓           ↓
                                     GbE
 SDH         SDH         PHY         PHY         PHY
  ↓           ↓           ↓           ↓           ↓
 WDM         WDM         WDM         WDM         WDM

  (a)         (b)         (c)         (d)         (e)
```

**Fig 6.12** Methods of supporting IP over WDM.

#### 6.7.1.1 *IP Over ATM Over SDH for WDM Transmission*

This quite expensive option is commonly used today because of the large base of ATM and SDH switches. IP packets are segmented into ATM cells and assigned different 'virtual connections' by the IP router. The ATM cells are then packed into an SDH frame, which can be sent either to an ATM switch or directly to a WDM transponder for transport over the optical layer.

One of the ways of trying to ensure a given IP QoS is to guarantee a fixed bandwidth between pairs of IP routers for each customer (layer-2 QoS management). ATM provides a way to do this with variable granularity by the permanent virtual channels (PVC) using either the ATM management system or switched virtual channels (SVC) dynamically set up within virtual paths (VP). It can also use statistical multiplexing to allow certain users to access extra bandwidth for short bursts from 1–100 Mbit/s. For IP traffic, which is by essence connectionless, the unspecified bit rate (UBR) traffic contract is mainly used within ATM networks. Nevertheless, if IP applications require a particular QoS, especially for real-time constraints, it is possible to use constant bit rate (CBR) or variable bit rate — real time (VBR-rt).

### 6.7.1.2 IP Over SDH, Packet Over SONET (POS)

This is commonly used to frame encapsulated IP packets for transmission over WDM (probably via transponders) using PPP encapsulation and HDLC framing for error control — also known as POS or packet over SONET. There are different types of IP over SDH interfaces:

- VC4 or concatenated VC4 'fat pipes' which provide aggregate bandwidth without any partitions between different IP services which may exist within the packet stream;

- channelised interfaces, where an STM16 optical output may contain 16 individual VC4s, with a possible service separation for each VC4 — the different VC4s can then also be routed by an SDH network to different destination routers.

### 6.7.1.3 IP Over ATM Directly on WDM

It is possible to have a scenario where ATM cells are transported directly on a WDM channel. The only difference is that ATM cells are not encapsulated into SDH frames, instead they are sent directly on the physical medium by using an ATM cell-based physical layer. 'Cell-based physical layer' is a relatively new technique for ATM transport. Some benefits of using a cell-based interface instead of SDH are:

- simple transmission technique for ATM cells since they are sent directly over the physical medium after scrambling;

- lower physical layer overhead (around 16 times lower than SDH);

- as ATM is asynchronous, there is no stringent timing mechanism to be put on the network.

### 6.7.1.4 IP Directly Over WDM

This method promises large cost reductions for transport of IP services. Using SDH interfaces on the IP routers means that in principle there is no need for SDH equipment. Various protocols (such as SDL) can be used to map IP directly over

WDM with less hardware. In the IP/WDM scenario, were the optical layer simple WDM point to point, then control and protection functionality would be implemented in IP layer 3. However, transporting IP over a true optical network means that some of this functionality will be achieved in layer 1, e.g. network protection.

### 6.7.1.5  IP Over Gigabit Ethernet for WDM

Ethernet accounts for over 85% of LANs worldwide. The new Gigabit Ethernet standard can be used to extend high-capacity LANs to MANs and maybe even WANs, using Gigabit line-cards on IP routers, which can cost about five times less than SDH line-cards with similar capacity. For this reason, Gigabit Ethernet could be a very attractive means to transport IP over 'metropolitan' WDM rings (Fig 6.13) or even over longer WDM links.

### 6.7.2  Protection and Restoration of IP Services

Three types of IP/WDM protection or restoration architectures can be identified to protect against failures in the optical or IP domains:

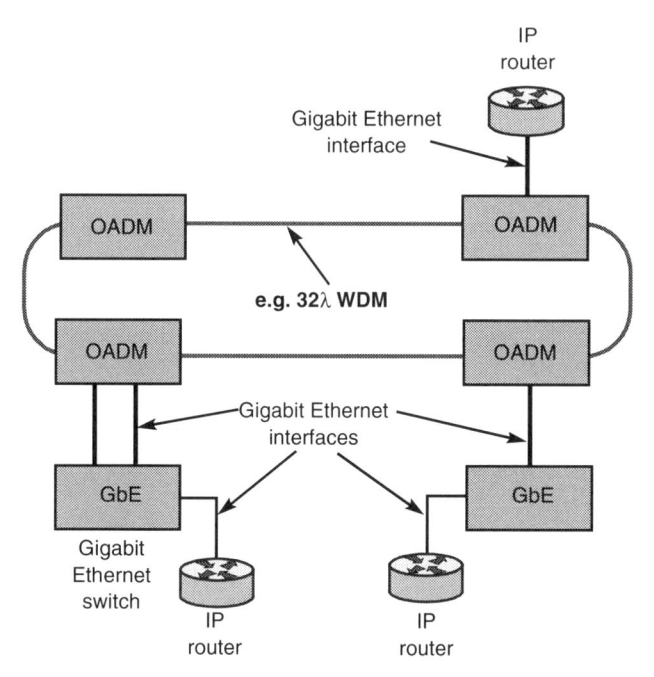

**Fig 6.13**  IP transported over a WDM metropolitan ring.

- protection against cable breaks using optical multiplex section protection (OMSP) restores a group of n wavelengths simultaneously using 1+1 protection or OMS-SPRing protection;
- protection against both transport equipment failure and cable breaks can be achieved using optical channel protection (OCHP) to restore wavelengths independently;
- protection against router failures is generally in the IP domain involving updating of routing tables, which may be as large as 64 Mbyte and may take several hours — other protocols such as MPLS will be able to update their routing/switching tables more quickly, especially for smaller networks, and a downtime of only a few seconds may be expected.

In addition, any combination of the above techniques can be used and Table 6.4 provides a summary of these protection options.

**Table 6.4**  Summary of protection options for IP services.

| Protection architecture | Degree of protection | Speed of protection | Cost | Remarks |
| --- | --- | --- | --- | --- |
| WDM OMSP | Fibre and optical amplifiers only | Fast, 50 ms | Lowest of all for high traffic | Fast and cheap, inflexible |
| WDM OCHP | Fibre, all WDM equipment | Fast, 50 ms | A little higher than OMSP | More flexible |
| SDH protection | Fibre, all SDH and WDM equipment | Fast, 50 ms | Depends on traffic volume, high for large traffic | Mature technology |
| IP restoration | All failures | A few seconds to hours | Needs larger routers Duplicated routers for high resilience | Needs careful IP dimensioning Slow restoration Synchronisation of routing tables |
| Combination | All failures | Fast, 50 ms for transport layer failures | Potentially highest | May be required for high QoS |

## 6.8  Developments in Metropolitan Networks

Metropolitan networks operate over relatively short distances using lower cost WDM technology (e.g. unamplified). The majority of traffic (say 70%) is routed and groomed within a region and typically has shorter spans, in the range 1–30 km.

A growing driver for metropolitan traffic is the development of eCommerce and data centres (Web hosting) where information can be stored, processed, cached and distributed within a small region. SDH will continue to be rolled out within metropolitan networks to support private services (with guaranteed QoS) and some data networks (Fig 6.14). However, SDH can be an expensive option for IP, ESCON and Gigabit Ethernet networks especially for shorter distances, and cheaper solutions include direct fibre drive, WDM point-to-point and optical networking.

Also, with metropolitan areas, the concept of dial-up wavelengths becomes very attractive for larger customers requiring large capacity on demand.

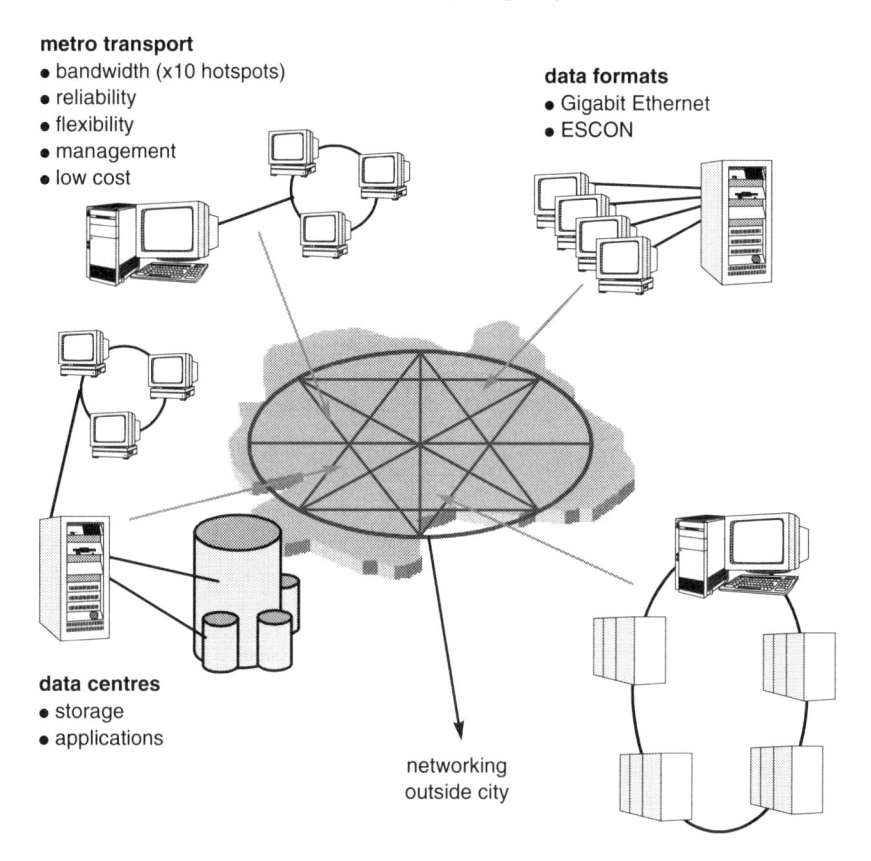

**Fig 6.14** Data networking in metropolitan areas.

## 6.9 Network Management

Having discussed network development and the technologies most likely to be introduced over the next few years into the transport network, this section now considers network management which determines the ability to make best use of the resources available. Typically releases of network management are phased in at about six-monthly intervals over several years following the deployment of the new network technology, so that the network eventually achieves full functionality.

84  *Network Management*

Networks are becoming increasingly complex both in their design and the services they must support. Increasingly operators are turning to their network management systems to solve these problems of complexity. The BT/UK core transport network is no exception — this section will look at how BT is managing its existing SDH network and how it is using the experience gained to shape its architecture for the future.

### 6.9.1 Current Architecture for Managing SDH

The BT systems architecture (Fig 6.15) has evolved over a number of years to be a large and complicated interconnection of many systems. There are three main threads to the system architecture. On the left, in Fig 6.15, are the systems involved with planning and building the network, in the centre are the systems involved with service provision, and on the right are the systems involved with service assurance.

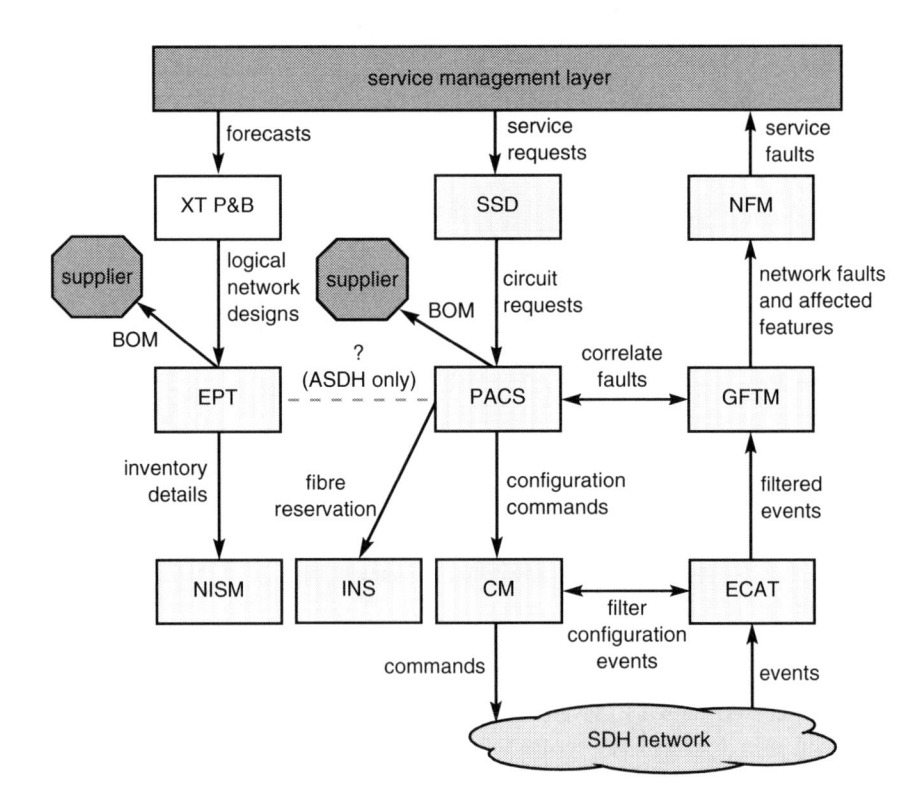

**Fig 6.15**  SDH network management system architecture.

The role of the systems involved in this architecture are briefly described below.

- Configuration manager (CM)

  CM carries out provision, cessation and rearrangement requests on behalf of PACS. It interfaces with suppliers' element managers to request the placement or deletion of individual cross-connections within the network elements being managed.

- Event collection and alarm translation (ECAT)

  ECAT polls suppliers' element managers for event information. It timestamps each event received and feeds them to GTFM.

- Equipment planning tool (EPT)

  Planners can use the EPT system to view existing floor plans and equipment layout, maintain equipment and generate the job pack for distribution to the field engineers. EPT provides visualisation tools for the equipment data held on both NISM and INS.

- Generic technology fault manager (GTFM)

  GTFM provides technology-specific fault management. This includes detailed filtration, association and correlation of event information received from the network.

- Integrated network system (INS)

  INS encompasses many subsystems supporting planning, assignment and maintenance of BT's PDH network. INS is also the master for fibre data and for location data.

- Network inventory and spares management (NISM)

  NISM is the physical inventory system for all technologies. It holds physical details of equipment and is also accessed by EPT to record planned equipment details. NISM is updated directly by the element managers when the planned equipment is installed.

- Network fault manager (NFM)

  NFM is responsible for bringing together fault information from across the BT network. These faults are associated together to help identify the root cause of the problem in the network. NFM also provides a company-wide trouble-ticketing system for handling problems.

- Planning, assignment and configuration system (PACS)

  PACS is the main system for the management of the SDH network. It supports the logical design of the SDH network and its physical realisation through the network planning and network build activities. It also handles all services

86  *Network Management*

requests providing a high degree of automation for providing, ceasing and rearranging circuits on the SDH network.

- Service solution design (SSD)

  SSD is responsible for designing circuits to meet customer services requests. This includes identifying and selecting appropriate technologies and designing the circuit routing across these different technologies.

### 6.9.2  Issues Arising from the Current Approach

At the heart of the SDH management architecture is PACS, designed and built in 1996 to handle the wide-scale introduction of SDH within the BT network. Its design is data centric with a single model of the network through which the functional modules control the operation of the network. The actual interaction with the network is carried out through the supplier element management systems. The single consistent view of the network provided through the data centric design has made it possible to develop highly automated business processes. Through the detailed data model, PACS has been able to handle the introduction of new network technologies such as SPRings without major redesign.

As the network evolves, so must the network management architecture, and two main development areas have been identified.

- Integrating technologies

  Partitioning planning around the network technology has lead to problems when trying to plan equipment space in a building and so the focus is now on combining the planning functions into a single set of systems regardless of network technology.

- Sub-network management

  Since the design of PACS equipment, suppliers have continued to improve their management solutions. The trend is towards systems that let the operator control the network on an 'end-to-end' basis as opposed to an 'element-by-element' basis. This means that an operator should be able to achieve similar levels of management control without having to develop such complex systems.

### 6.9.3  Improving the Current Architecture

The potential introduction of new networking technologies into the BT/UK core network affords the opportunity to address some areas of the current architecture in order to:

- enable the BT/UK network to be planned as a whole — regardless of technology;
- provide the opportunity to trade flexibility for speed of response;
- recognise that equipment vendors should be in the best position to manage the detailed workings of their own equipment.

A key change would be to separate out the planning functionality from the 'real-time' aspects of managing the network such as provision of service and service assurance, as shown in Fig 6.16. These real-time functions are supported by the domain management systems (DMSs) or the cross-domain management system (XDMS). The architecture recognises that the existing network must continue to be supported, but with the focus moving towards the new high-bandwidth networks — this support is limited to continuing to manage these networks in much the same way as before.

It is clear that network operators can no longer afford the luxury of building bespoke management solutions and must look to the equipment suppliers to provide at least part of the solution; however, the solutions offered must be capable of meeting the increasingly sophisticated needs of the operators.

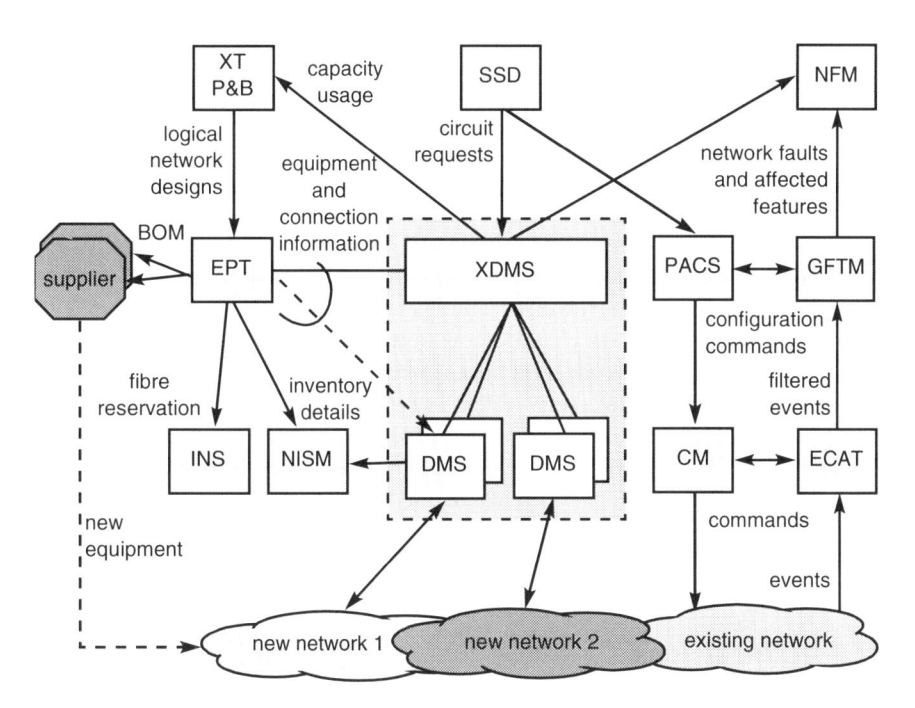

**Fig 6.16** A new management architecture.

## 6.10 Future Network Management Issues

The convergence of the optical and packet networks and the increased drive for network efficiency is likely to have a significant effect on the way the networks of the future are managed. Currently, to meet an increased traffic demand in the packet layer might take days if not weeks to resolve through traditional processes. This process would typically involve different parts of the organisation using different management systems working together to identify where the problem lay and the best course of action. If the solution is, as is often the case, the provision of additional network capacity, this can increase still further the time to respond. If the case is also considered where there might be a need to remove excess capacity from the network, the whole process starts to look very inefficient both from a network and a business perspective.

What is needed is a management approach that closely mirrors the dependencies between the layers in the network. Such an approach could be achieved by embedding more of the management functionality into the network. This is not a new approach. No one today would expect to reconfigure an SDH ring to restore traffic following a fibre failure — the problem is detected and handled by the network elements following a set of rules or policies that define its operation under failure conditions.

The future could see this same approach applied between layers in the network. To return to the previous example, if having detect the congestion in the packet layer the network was able to automatically provision additional capacity in the optical layer to meet this need, the problem could be resolved within hours, not days or weeks as before. This speed of response and level of automation would make it possible for the network to start to offer a degree of self-optimisation.

To control this increased level of intelligence the management system must define the bounds within which the network can exercise its intelligence. This would lead to a management style that approaches the way IP networks are controlled, where the management system is responsible for setting policy rather than explicitly defining the use of resource. It is envisaged that this approach would lead to a much more dynamic overall network, but this is not without its problems — it will become increasingly difficult to predict how the overall network will behave over time, which will no doubt give rise to a whole new set of management applications.

## 6.11 Conclusions

This chapter has examined the drivers and technologies which will influence transport networks over the next five years.

The detailed decisions on which technologies and management systems to deploy will be taken jointly with BT suppliers, but it has been possible here to describe the current mainstream technologies to support the private services platform and the emerging data platforms. It is also recognised that network management must evolve into new systems architectures to achieve scalability as the BT core transport network develops into the future.

## Reference

1  Hawker, I.: '*Future transport networks*', British Telecommunications Eng J (July 2000).

# 7

# DELIVERY OF IP OVER BROADBAND ACCESS TECHNOLOGIES

## M Enrico, N Billington, J Kelly and G Young

## 7.1 Introduction

The start of the new millennium coincides with increasing global deployment of broadband access technologies, many of which are being used to deliver IP-based services and applications to customers at increasingly faster speeds.

This chapter provides an overview of how to supply broadband access, i.e. high-speed Internet services, to the population at large. DSL and cable modems are currently being deployed across the globe. LMDS radio licences have been issued in several countries now and satellite technology has transitioned from a focus on the mobile, narrowband, backhaul and rural markets to being looked at as a viable broadband access system. In addition, an increasing amount of optical fibre is being deployed in access networks. Metropolitan fibre rings from a range of different carriers are now being constructed in many of the developed world's largest cities.

For IP, the trend towards broadband access opens up new service opportunities facilitated by the increase in bandwidth. High-quality streaming entertainment video, videoconferencing and interactive gaming are just some of the possibilities where the removal of the access bottle-neck significantly improves the customer's experience of the service. In addition, broadband access facilitates the bundling of multiple services simultaneously through the same access 'pipe'. This opportunity also brings with it new challenges for IP and its supporting network infrastructure and systems. The ability to prioritise certain traffic types over others and incorporate quality-of-service guarantees will be key network differentiators in the new competitive broadband era. Business models will evolve from the simple 'connect to an ISP to download information' approach. Increased functionality at local access nodes (caching, switching, routing) may be required of some of the new business models and applications. Broadband access technologies are set to compete with and complement each other as operators race to build the 'best integrated IP network'.

There is great interest, not just from ISPs and network equipment vendors but from politicians and the media, in the future of broadband access technology as this is seen as a key enabler for the future of the Internet. The chapter overviews digital subscriber line technology, cable modems, wireless local multipoint distribution systems and satellite technologies. It will allow readers to understand the terminology and help to explain how to use the technologies in real network configurations.

Firstly, a brief overview is provided of some of the key broadband access technologies that are being used to address the residential and SME markets. The focus then switches to describe BT's approach to wireline broadband access using ADSL, the delivery of current broadband IP services over this platform, and the potential evolution of the wireline access network platform.

## 7.2    A Brief Comparison of Broadband Access Alternatives

This section gives an overview of the ADSL, cable modem, LMDS and satellite technologies for broadband access (see Fig 7.1). It highlights the salient features together with the strengths and weaknesses of each.

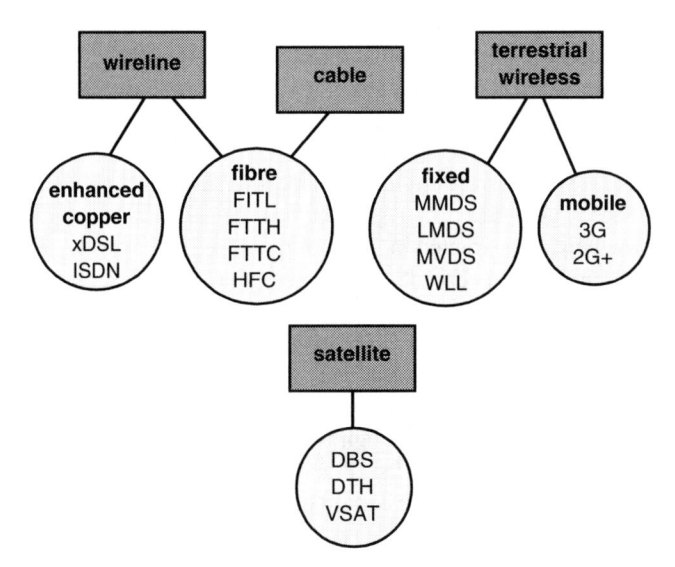

**Fig 7.1**   Broadband access technologies.

### 7.2.1 ADSL

Asymmetric digital subscriber line (ADSL) is essentially a very high-speed modem enabling transmission of megabit rates over existing copper telephone lines. Unlike voiceband modems, it can provide the data transmission capability on the line at the same time as the existing analogue telephony service [1] as shown in Fig 7.2. A major strength of this technology is that there are over 700 million copper telephone lines in the world. Most customers therefore already have the necessary bearer infrastructure connected to their premises. Capital expenditure to create an ADSL line is matched by a potential new revenue stream. Thus ADSL reduces up-front speculative expenditure by deferring customer equipment deployment until the customer requests service, i.e. just-in-time (JIT) provision of service with reduced sensitivity to service take-up.

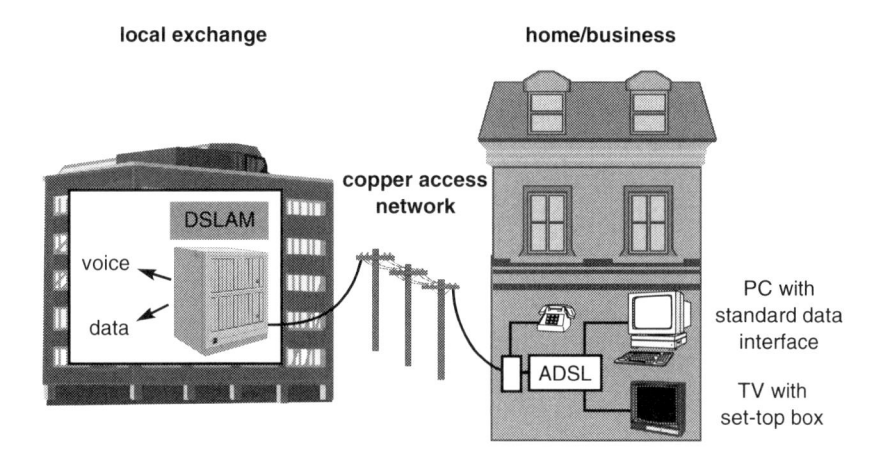

**Fig 7.2** Broadband access over ADSL.

An individual copper pair is a point-to-point connection dedicated to a customer, thus making it possible to offer guaranteed performance bounds and a secure access connection.

A weakness of DSL technology is that each individual customer's copper line is different (in terms of length). Hence for some service offerings a line qualification process may be required to determine exactly what bit rate the customer's line can support. Not all customers can receive the fastest rates which could vary from 8 Mbit/s to customers close to an exchange to 1 Mbit/s for a customer on a longer rural line.

With the advent, in some countries, of local loop unbundling (LLU), operators other than the incumbent can connect their own DSL systems to the copper access network. Crosstalk between different DSL systems in the cable causes noise and interaction between adjacent systems. This has been well controlled in the hitherto single incumbent operator environment.

However, in the new era of LLU, agreement among operators on which systems can be safely deployed in the network and at what locations becomes necessary in order to avoid spectral pollution which could compromise network capacity for all users. Policing the deployment of equipment for mutual compatibility and identifying the source of service problems becomes a significant technical and administrative challenge.

### 7.2.2 Cable Modems Over HFC Networks

The main advantage of cable networks over other broadband access networks such as ADSL, LMDS or satellite is the large amount of per-customer bandwidth available over the access transmission media. However, much of the downstream bandwidth is used for TV and, in general, the upstream bandwidth is limited both in terms of the spectrum available and by noise. The total bandwidth of such a system can be 860 MHz [2] or even approaching 1 GHz. This allows the use of frequency division multiplexing to mix a large number of services and channels on a single broadband pipe into the home or business. This gives operators a great deal of flexibility in packaging service bundles for a 'one-stop shop'. Old cable networks used a large amount of coaxial cable (in a tree-and-branch topology) with the associated need for many amplifiers. Modern networks are hybrid fibre coax (HFC) with an increasingly high fibre content, perhaps to within a few hundred metres of customers' homes (see Fig 7.3). The combination of deeper fibre penetration in the cable access network combined with modern digital modulation techniques has increased the bandwidth that can be delivered to cable customers.

Cable modems operate by using unused TV channels in the downstream direction to the customer and the lower frequency reverse path bandwidth for the upstream or return direction. A downstream bit rate of 30–40 Mbit/s per 8-MHz channel slot is possible. This capacity is shared across a number of customers[1]. The total aggregate upstream data rate is in the order of 10 Mbit/s shared across 50–2500 customers depending on the network topology. Most customers' cable modem units will not transmit at much more than 1 Mbit/s upstream. The downstream channel is continuous, but divided into cells or packets, with addresses in each packet determining who actually receives a particular packet. The upstream channel has a media access control that slots user packets or cells into a single channel.

---

[1] Note that no single customer could receive this full rate since the 10 Base-T Ethernet card in the PC would limit the burst speed to less than 10 Mbit/s.

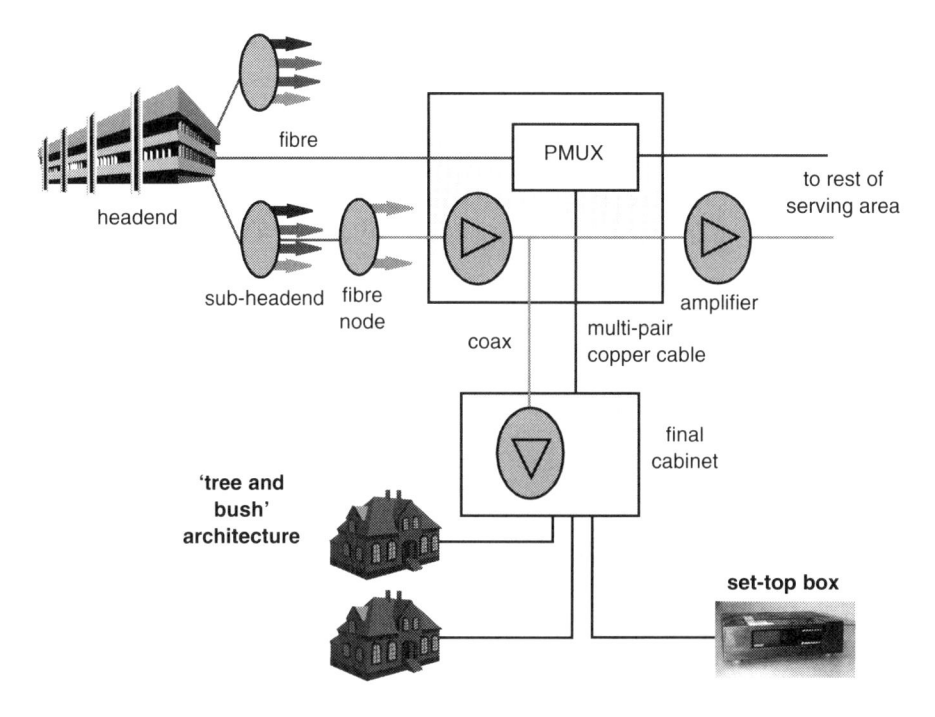

*Delivery of IP Over Broadband Access Technologies* 95

**Fig 7.3** A typical modern cable (HFC) network.

The cable network is broadcast in nature with the head-end cable modem broadcasting to all customers on the cable network that can receive the 8-MHz channel containing the digitally encoded data. It is rather like a large Ethernet LAN spread across the locality. This has two major impacts on performance. On the positive side this means that there is efficient statistical multiplexing of capacity across active users. An individual customer's modem can burst up to very high rates. The downside is that during the 'busy hour' when many users are surfing the Internet, average throughput will be degraded. Hence performance is unpredictable and careful network dimensioning is needed to give a good experience to the end customer. Quality of service will also degrade as Internet users on a network shift from text and simple graphics to high-quality graphics and multimedia, an inevitable trend if the Internet is in any way successful. The 'best-effort' nature of existing cable modems affects their ability to effectively support delay-sensitive services such as videoconferencing. It is possible to use the cable modem management system to impose limits on traffic sent downstream and to throttle back users trying to take more than a fair share of the capacity. Such management software may also be used to introduce a degree of QoS.

The broadcast nature of the cable modem means that each customer modem unit has to identify which IP packets are destined for itself and ignore the rest. This has led to security breaches where hackers have examined other people's traffic in the neighbourhood and also accessed other customers' PCs. These types of problem may be overcome using encryption and firewall techniques if the user is sophisticated enough to implement them and willing to pay for them. The improved DOCSIS 1.1 cable modem standard is designed to overcome such problems. Within Europe, there has not been widespread adoption of a single cable modem standard giving rise to a potential 'battle' between DVB/DAVIC and EuroDOCSIS standards.

### 7.2.3 LMDS

Local multipoint distribution system (LMDS) is a radio technology that provides broadband network access to many customers from a single base-station [3] as shown in Fig 7.4. Although LMDS specifically refers to a frequency allocation in the United States, the term is generically used to refer to broadband, multiservice radio access systems. These systems are also sometimes known by other terms, such as broadband wireless access (BWA), broadband point-to-multipoint (PMP), or broadband wireless local loop (B-WLL). A major benefit of broadband radio access is that once the base-station is in place, the remaining infrastructure required is only the customer units. Hence this enables provision of high-bandwidth access in a very expedient manner. LMDS can be used to extend the coverage of fibre rings without the need to negotiate way-leaves and build infrastructure, such as cable ducts (although it is usually necessary to negotiate roof rights for both base-station equipment and customer equipment).

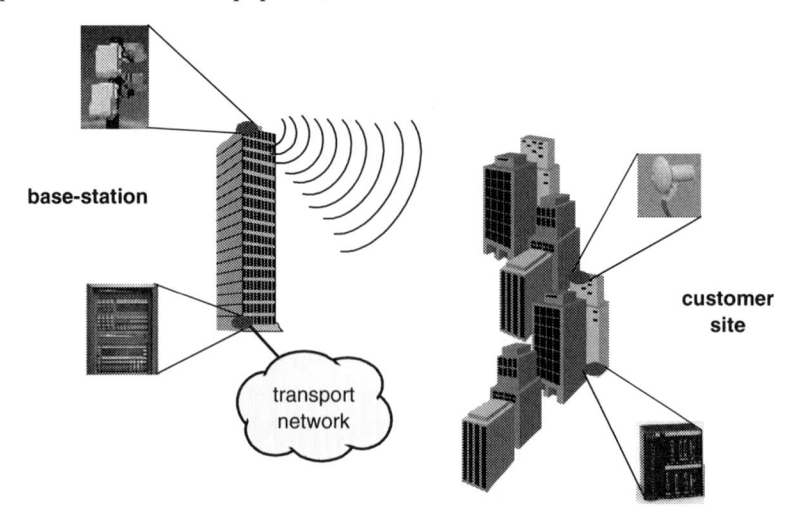

**Fig 7.4** Broadband access using LMDS.

Customers are connected within a 'cell' that typically has a radius of around 2 km from the central base-station. The base-station is usually connected to the remainder of the network using fibre or point-to-point radio. The base-station acts as a hub for the network and provides service to customers who are in direct line of sight and within the cell radius. Generally, the cell area is split into a number of sectors which allows the cell radius, and the capacity offered within a cell, to be increased. LMDS allows flexibility in the way that capacity is allocated to customers, e.g. asymmetric circuits can be allocated in addition to symmetric circuits and quality-of-service options can be offered for data circuits. Many systems also allow bandwidth to be allocated on demand.

The point-to-multipoint topology of an LMDS system is not dissimilar to that of a cable modem in that the base-station sends information to all end customers within a radio sector on a single radio link and the customer premises equipment selects the information intended for it. Typically a base-station may provide a symmetric capacity of 500 Mbit/s to be shared between the users in a cell. The per-customer allocation may be around 7 Mbit/s symmetric, which can be further shared from the customer outstation (via multiple ports) between a number of offices or desks. LMDS can outperform ADSL in terms of upstream data rates. The core network infrastructure used to connect to LMDS base-stations is virtually the same as that used to connect ADSL exchange units (DSLAMs) and hence the technologies can be deployed in a complementary manner.

The radio spectrum allocated to access systems (usually around 10, 26, 28 or 38 GHz) is finite and therefore some degree of planning of the radio resource is required. The limited radio spectrum means that LMDS is better suited to bursty data than real-time high-quality video applications. That having been said, the support for real-time applications is probably better than on cable modems; however, there is not generally enough upstream bandwidth for this support to be comparable to that offered by ADSL. Another point to note is that, as LMDS systems tend to be based on ATM, their QoS support is not so immature.

One of the key limitations with LMDS is the requirement that there is a line-of-sight path between the customer and the base-station. The availability of this path is obviously dependent on the geography of any city where a network is being deployed. To ease the line-of-sight problem, it is usual to mount the base-station on a building with good visibility over the surrounding land (requiring negotiation with landlords). Millimetre wave systems are affected by rain and snow as these cause the signal to be attenuated leading to an increased bit error rate on a link. However, the impact of rain is well understood and can be accounted for by appropriate link margins in the network design. Typically, line-of-sight and interference considerations mean that only 50–70% of the potential customers in the nominal 2-km cell radius around a base-station can be provided with service. However, it should be noted that a few customers out to 4.5 km may still be reachable with good availability (the actual availability is highly system specific).

One of the factors that has the greatest impact on whether an operator can deploy LMDS is the availability of radio spectrum. Since radio spectrum is generally licensed by government agencies, an operator will have to bid for a licence before being able to use the technology. In Europe, the main band of interest is the 26 GHz allocation, although 28 GHz, which is used in the USA, is also being proposed in some countries. The UK is also considering 40 GHz. Due to this lack of global standardisation, the exact frequencies associated with a spectrum allocation may differ from country to country. In some countries, licensees may be obliged to provide coverage of a certain minimum area or certain number of base-stations within a time limit. The cost of the licence could affect the viability of using LMDS in certain geographies. In addition, the customer units are currently priced higher than equivalent DSL or cable modem units. Therefore use of LMDS has been more focused on the medium-size business market and not the residential market. Distribution of broadband to concentrations of customers in clusters (such as shared office buildings and business parks) has been used as a way of sharing the common costs among a number of users to reduce the cost per customer.

### 7.2.4 Satellite

Satellite is used to deliver broadcast entertainment TV to many consumers around the globe. More recently this 'downstream' broadband capability has been combined with narrowband terrestrial return paths (modem or ISDN) in order to add an interactive capability to services (see Fig 7.5) [4]. Two-way satellite systems

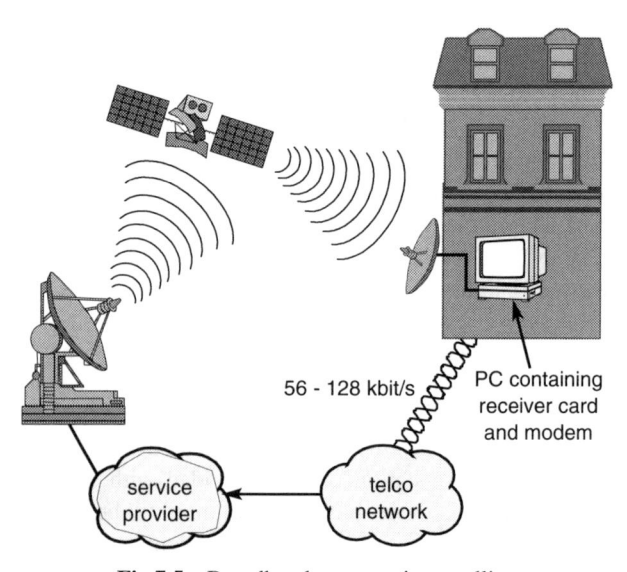

**Fig 7.5** Broadband access using satellite.

using very small aperture terminals (VSATs) have also been used by business customers. Assuming access to a satellite transponder is established, satellite systems can then be used to provision service rapidly across a wide geographic footprint to provide access to a large number of distributed customers. Since satellite was originally designed for broadcast, it has an inherent capability to support multicast services and 'push' technology. In an IP network, satellite can be used to update local caches and to deliver streaming IP video. A key strength is the ability to provide ubiquitous access at a consistent quality level across a large area. The cost of deployment is independent of geographic distance within the satellite footprint. Satellite access is less affected by local terrain than many alternative broadband technologies. This makes it an ideal technology to use in rural areas where it has the potential to be more cost effective than the alternative wireline or terrestrial wireless systems. Satellite technology has been used for a while to provide business TV services to corporate sites. This can now be combined with IP in order to link this capability into Intranets. For example, streaming IP video or software upgrades could be delivered to a site via satellite and then distributed to users' desktop PCs using the existing IP intranet over the site LAN.

## 7.3  Near-Term Evolution of BT's Broadband Platform

In the near term, BT's broadband platform is likely to evolve towards a feature-rich, multiservice delivery platform while at the same time supporting simple, low-cost, high-speed Internet delivery. ADSL network termination equipment (often referred to as the ADSL modem) will become CPE and be owned by the customer. Innovation in the CPE arena will place new requirements upon the network to which it is connected, while at the other end of the spectrum the desire for cheap, fast Internet connectivity will be addressed by simple ADSL modems and corresponding basic service offerings.

### 7.3.1  Multiservice Delivery Over ADSL

Innovation in the CPE arena is leading towards the ADSL modem becoming a portal into the home for a range of different services which is attractive to the customer by facilitating a one-stop-shop approach to communications into their home. However, bundling services to include a mix of video, voice and data communications requires an evolution in the capabilities of the access delivery mechanism. Differing services may place differing and possibly conflicting requirements on the access delivery mechanism. Streamed video is characterised by being tolerant of delay but very intolerant of transmission errors. Voice requires low latency, and is relatively tolerant of errors. Data is generically tolerant of both delay and transmission errors; however, in the context of TCP/IP the delay must be minimised in order to prevent a knock-on effect on data throughput.

ADSL was originally conceived as a delivery mechanism for streamed, compressed video over the telephone line access network. The high tolerance of video to delay meant that ADSL was able to employ physical layer forward error correction and interleaving techniques to mitigate the effects of impulse noise on the integrity of the bit stream. Interleaving, in particular, is critical in the defence against impulse noise events commonly encountered in the metallic access network. However, its downside is that it introduces significant delay. Hence, for ADSL delivery, low delay is synonymous with burst error events, while, if higher delay can be tolerated, virtually error-free transmission can be achieved. Typical round-trip delay figures for ITU-T compliant ADSL are up to 44 ms if interleaving is employed and 5 ms if it is disabled. When considering transmission of multiple, mixed services over an ADSL access system, the link between interleaving depth (i.e. delay) and error rate leads to the logical conclusion that different interleaving depths may be required for different services. ITU-T ADSL standards allow for two different interleave 'paths' through the ADSL link whereby the data stream can be split between a 'fast path' and an 'interleaved path', a concept referred to as 'dual latency'. However, dual latency is optional and is not commonly implemented. Furthermore it is not possible to vary dynamically the share of the available bandwidth between the two paths — hence the bandwidth allocation is usually fixed under management control. This leads to limitations for multiple service delivery where it may be desired to share ADSL line transmission capacity dynamically between more than one service with differing latency requirements. One possible way round this dual latency problem may be to carry all traffic over the same low-latency (non-interleaved) ADSL path and employ appropriate transport layer protocols (e.g. TCP) to ensure error-free delivery for error-sensitive applications. However, this is not appropriate for real-time services that cannot tolerate any delay due to retransmission and buffering.

Using ADSL to carry multiple simultaneous services also requires some ability in the DSL access multiplexer (DSLAM) to allocate traffic resources according to the required quality of service. Many implementations to date have relied on simple connection admission control rules at configuration time to allocate resources. However, where the mix involves some guaranteed (CBR) type traffic and some best-effort (UBR) type, then multiple QoS queues together with appropriate methods of dealing with excess traffic (e.g. EPD, PPD, weighted fair queuing, traffic shaping) will be required. DSLAM implementations are becoming more sophisticated in this area and such features will become a core requirement for reliable multiservice delivery.

An implication of using a single ADSL line to carry more than one service is that the transmission bandwidth requirement is likely to exceed that of the single service case. ADSL, like any other transmission system, is bound by Shannon's capacity theorem which defines the maximum transmission capacity for a given bandwidth and signal-to-noise ratio. In order to be able to guarantee ADSL service on a given line, a pessimistic assumption of crosstalk noise environment is made which holds true in all but the most extreme cases. From this assumption, for any given transmission bandwidth requirement, a corresponding maximum length of line can be defined. Figure 7.6 shows a typical relationship between ADSL transport capacity and line loss for 0.5 mm copper cables in the BT access environment. Thus as the transmission capacity requirement rises, the percentage of the customer base which can be served by ADSL reduces. ADSL is, by definition, an asymmetric transmission system (i.e. capacity in one direction is lower than in the other). The original requirements for ADSL, and the environment in which it has to operate, result in an asymmetry ratio of approximately 10:1 downstream to upstream, which is fixed for standards-compliant ADSL equipment. However, such an asymmetry ratio is not ideal for all services.

ADSL is only one example of a family of xDSL transmission systems and others are able to exploit degrees of freedom not available for ADSL. For example SDSL (symmetric DSL) is optimised for symmetric operation over medium line lengths and IDSL (ISDN DSL) is optimised for low symmetric bit rates over long line lengths. There is now an increasing trend for manufacturers to produce universal DSLAMs with the capability to support a variety of different line cards in the xDSL family. This will give operators the flexibility to optimise network penetration for all services based on optimal choice of line transmission technology.

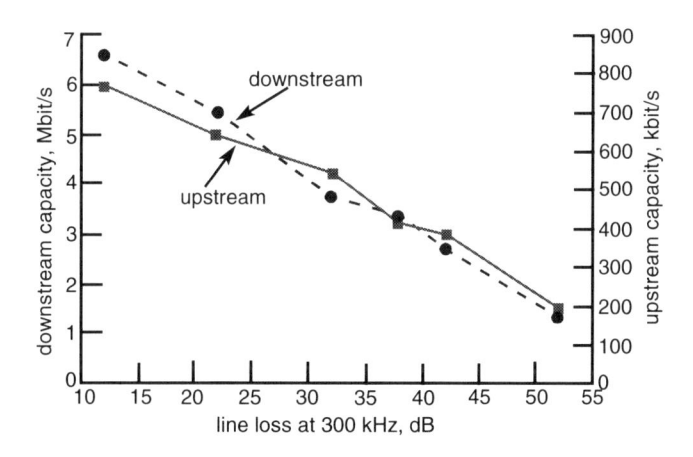

**Fig 7.6** ADSL capacity versus line loss.

### 7.3.2  Basic ADSL for Internet Access

At the opposite end of the scale, the requirements of the single PC Internet surfer are considerably simpler. This type of user is not reliant (within reason) on a minimum access transmission rate, but delay is a key parameter. The interaction between the

TCP window mechanism and the delay/bandwidth product of the communications link can limit the speed of data transmission if there is significant latency. This leads to operation of ADSL in 'fast' mode (i.e. non-interleaved) being popular for Internet connectivity. The resulting susceptibility to occasional error bursts can be taken care of by the TCP retransmission scheme.

The lack of a hard minimum bit rate requirement, means that the ADSL link could be operated on a best-effort basis in this case. Service would then follow the dial-up modem analogy where the bit rate is not guaranteed. Such an approach would significantly increase the potential penetration of ADSL, because those with longer lines would continue to work albeit at a lower bit rate. It is the upstream (i.e. customer-to-exchange) direction which usually limits the line length because of the conflict of service requirements with the limitations of the upstream portion of the copper-pair transmission channel. Hence reducing the upstream ADSL bit rate typically has the most effect on increasing the range. A best-effort ADSL service would also negate the need for sophisticated line pre-qualification, which is required for fixed bit rate services to see if the customer is within range. The bit rate supported by an individual line would vary according to the amount of other xDSL crosstalk in the cable. This would be likely to get worse with the increasing deployment of xDSL and therefore a best-effort ADSL service could be expected to degrade with time. The range of speeds at which an ADSL modem could run in this mode is significant and it is difficult to imagine how such a service could be billed, with some customers getting much faster service than others when subscribing to the same service. Billing per megabyte is one possible solution but is fraught with difficulties, especially with the advent of push technology where customers may not explicitly have requested all the data which they end up receiving.

G.lite has received considerable press attention in recent months as the answer to cheap, high-speed Internet access using ADSL transport. It was conceived as a variant of ADSL which could be easily installed by the end user by plugging into a phone socket. However, problems with voice interference have been encountered during use and some form of filtering is commonly implemented in practice (see below). Functionally, G.lite uses a cut-down ADSL modem function which has a maximum transmission rate capability of 1.5 Mbit/s downstream and 500 kbit/s upstream. Power consumption and chip area cost savings are possible over normal ADSL. A major advantage of the technology could be that exchange power consumption and hence line-card packing density could be significantly improved. However, operators typically wish to offer a range of services over ADSL, not all of which are within the capabilities of G.lite, and hence exchange deployments are typically of dual-mode line cards which can support either standard (G.DMT) ADSL or G.lite, where the G.lite power saving cannot be realised. G.lite may find its niche where size and power consumption are key drivers, e.g. where the line card is housed within remote street electronics or in CPE, such as laptop PCMCIA cards.

### 7.3.3 The ADSL Modem as CPE

There is a strong commercial case for moving away from the traditional view that the customer-end ADSL equipment is network termination equipment (NTE) which is owned, installed, managed and maintained by the operator. However, from a technical viewpoint this raises a number of challenges:

- the baseband PSTN service must pass through a filter associated with the ADSL NTE in order that the ADSL and PSTN signals can be separated,

- there is still a variation in the quality of ADSL equipment from different manufacturers — the quality of the equipment in this context will determine the maximum length of line over which it will work and hence will affect the ability of the operator to predict if the service is suitable for a given line,

- the higher layer functionality of the ADSL NTE is closely linked to the requirements of the service — innovation in the NTE arena will require corresponding innovation in network services (functions above basic ATM transport are not covered by ITU ADSL standards).

The POTS splitter filter has traditionally been included in the ADSL NTE or been closely associated with it. The specification of this filter is critical if good voice quality is to be maintained on the baseband PSTN service. BT's POTS filter requirements, in common with many other European operators, are far more stringent than those of the United States because a complex telephone line terminating impedance is used. There are two main candidate options for installation topologies which enable the customer to install their own ADSL modem — distributed micro filters and a BT-owned network terminating device containing the low-pass portion of the splitter.

The distributed micro-filter topology has been adopted as the default method for installation of G.lite or 'splitterless' ADSL. Originally designed for operation without a POTS filter, practical experience of G.lite has indicated that DSL interference with the voice service is significantly improved if a filter is applied to each telephone lead (Fig 7.7).

Other advantages of this approach are that the installation is entirely carried out by the customer, and that a telephone socket convenient for the location of the modem can be used to connect it to the network. On the other hand, design constraints on the micro filters result in significant degradation in the quality of the voice service.

This manifests itself as increased side tone which is unpleasant to the user and can make the volume of speech from the other end seem much quieter depending on ambient room noise.

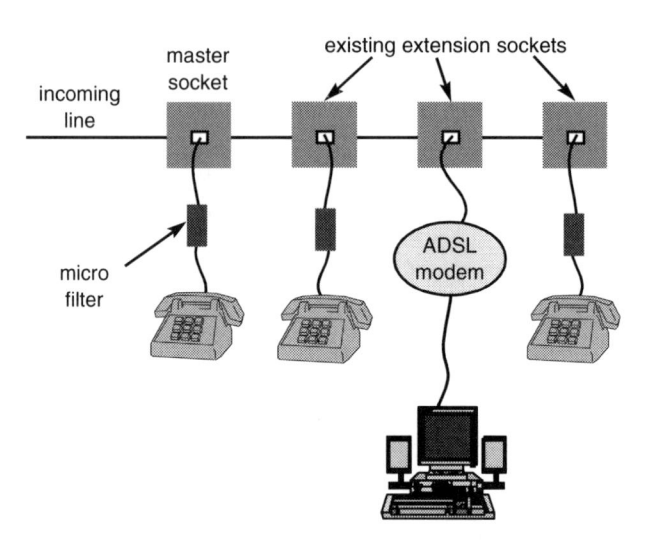

104  *Near-Term Evolution of BT's Broadband Platform*

**Fig 7.7**  Distributed micro filters.

BT's initial approach for launch of ADSL-delivered services will be to fit a line termination device containing the low-pass portion of the splitter (Fig 7.8). This will be realised as a replacement plug-in front plate for the NTE5 PSTN master socket. There will then be two sockets on the front of the master socket — one phone socket and one for connection of the ADSL modem. The disadvantage of this approach is

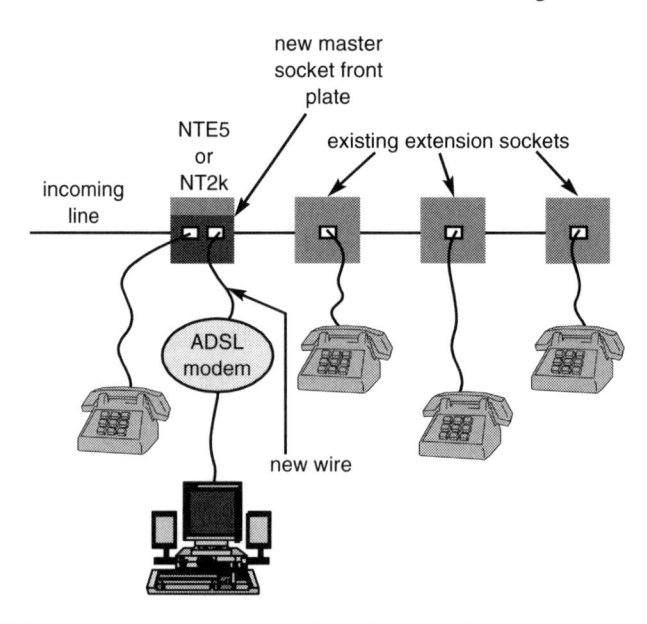

**Fig 7.8**  Topology of a customer-end installation with a master socket filter.

that BT will initially have to install the new line termination device and that an extension cable will be required from there to the desired location of the modem. On the other hand, the filter can be designed to have minimal impact on voice quality while still enabling the customer to connect the ADSL modem themselves. Longer term, in preparation for the ADSL modem becoming CPE, it is intended that electronics will be included in the network termination box allowing better line diagnostics and fault demarcation to be carried out. The new terminating box is known as the 'NTE2000'.

The issue of variability of quality in ADSL modem design is being addressed by industry and standards bodies. The DSL Forum has facilitated a series of industry 'plug fests' where vendors can meet and carry out interoperability experiments on neutral ground. The recently published ITU standards for ADSL include performance requirements for each major region of the world. Compliance with these standards should ensure that baseline performance targets are met.

Higher layer protocol requirements of services carried over ADSL are currently less well defined. Basic ATM and packet transport over ADSL recommendations have been produced by the DSL Forum, and ITU ADSL standards cover the transport of basic ATM cell streams. Innovation in CPE and services, will require the definition of significantly more functionality than is covered by existing standards. In the near term it is likely to fall to the operator to publish interface specifications in order to define the requirements of specific services. In the UK these interface specifications are referred to as supplier information notes (SINs).

## 7.4 The Roles of IP and ATM in Broadband Access Systems

The history of several years of silicon and systems development of today's broadband access technologies has contributed to the IP versus ATM debate of recent years. In some parts of an end-to-end network, bandwidth can be used to solve problems associated with quality of service. For example, DWDM can be used in the core and Gigabit Ethernet in the customer's building. However, apart from using fibre or coax bearers, most access delivery media do not have the luxury of excess bandwidth. When today's broadband access systems and silicon for ADSL, LMDS and satellite were first being developed several years ago, ATM was the only way of managing and policing traffic and offering absolute QoS mechanisms that could underpin service level guarantees. Hence use of ATM was subsequently embedded in ADSL silicon and IP was seen more like 'just another application' to be transported rather than a comprehensive networking technology. The LMDS air interface is based on ATM to enable statistical multiplexing and assignment of bandwidth on demand. It is only recently that IP developments, such as RSVP, Diffserv and MPLS, have come along to move IP-centric networks towards equivalent functionality. The use of ATM in much of today's DSL silicon is, of course, not to the exclusion of IP. In the DSL Forum's system guidelines for ATM-centric architectures (as opposed to its packet-based recommendations), IP is carried over ATM (as described in the next section).

Several network operators are taking advantage of this IP over ATM approach by using an end-to-end architecture enabling product offerings for either layer-3 IP or layer-2 ATM services over a common platform. In fact in some parts of the world the regulators give the operators little choice but to offer a wholesale layer-2 service, enabling service providers to add their own IP layer. Hence their service provider customers then have the choice of which network product best suits their services. For instance, some companies procure the IP product to construct IP VPNs and others procure the ATM product, e.g. for VoD. The ATM products are not restricted to larger traditional operators. In the USA, new network operators (CLECs) which have only been in business a couple of years are providing their own DSL equipment on unbundled copper loops rented from the incumbent telco. These companies were starting with a blank sheet of paper when it came to designing their network. Some of them are offering only layer-2 ATM network transport as a wholesale product because that is what their ISP and corporate customers require.

When the Universal ADSL Working Group (UAWG) was formed, its focus was a mass-market consumerisation of ADSL and it proceeded to produce the specification for DSL-lite (G.lite). This initiative was driven largely by the PC industry. The existing ADSL specifications were used as the starting basis for DSL-lite in UAWG work and ITU standards. Given that ADSL silicon had taken several years for the leading vendors to produce their most integrated chips, the UAWG also followed the IP-over-ATM approach in order to expedite time to market. Consequently the bundling of DSL-lite into PCs via mass-market distribution channels also enables both ATM and IP services to be transported.

One of the most recent broadband access developments is that of voice over DSL (VoDSL). There are two ways that the voice can be carried across the ADSL — VoIP or VToA. Again industry is split and is pursuing both approaches since there are perceived to be two very different markets for VoDSL. Broadly speaking one market is addressed by completely replicating the quality and feature functionality of today's PSTN by using voice over ATM. The other is to dispense with the legacy baggage of today's PSTN and to focus on new value-add services and integration with data (e.g. click-to-talk Web sites using VoIP) without replicating the exact quality levels of the PSTN. The development of VoDSL enables voice and data services to be simultaneously delivered down a single copper telephone line. This bundling of digitally derived voice together with data will be equally applicable to other broadband access technologies where VoIP and VToA could be options.

There is logic (technical and marketing) in the history behind why ATM functionality has ended up in broadband access silicon and systems and there appears to be very real markets for both ATM and IP delivery over such systems today. However, the more important issue is what will the markets require tomorrow (e.g. multicast) and how should the technology evolve to best serve those markets? The world is moving towards an increasingly IP-centric future. It seems less likely that the ATM layer will be removed from, say, ADSL since the impact on silicon and interoperability of mass-market ADSL/DSL-lite products would slow down broadband access to the masses. What is possible is that, as the new developments for IP QoS and connection-oriented capability, etc, are developed in the Internet Engineering Task Force (IETF), the ATM layer within broadband access systems may then become 'dumbed down' so as not to duplicate addressing and signalling mechanisms at two layers. The relative merits of ATM or IP-centric broadband access systems will continue to be debated, with vendors pursuing and further developing their own preferred approaches. Already DSLAMs have been evolved from simple VC cross-connects. Some vendors have developed DSLAMs that are SVC-capable ATM edge switches and others have developed them with integrated IP routing and multicast capability. Proprietary LMDS systems are also now starting to appear with more inherent IP capability. Multiprotocol label switching (MPLS) is seen by some vendors as the way to integrate IP and ATM capability to get the best from each. Its role in broadband access systems is yet to be well defined but some vendors are pursuing it. As always, interoperable standardised products are preferred by many operators and progress in this area could dictate the speed of adoption and ultimate success in the market of the ATM, IP and MPLS approaches to broadband access.

## 7.5  Delivery of IP Services Over BT's Broadband Platform

The possibility of 'always-on', mass-market, broadband IP services has had a significant impact on how ADSL network operators design their broadband IP network platforms. Many factors have had to be taken into account, the most important of which are as follows:

- an 'always-on' service in the strictest sense of the word implies that the IP addressing scheme has to be static, which is virtually impossible to achieve for mass-market broadband IP services given that public IPv4 addresses (administered by RIPE[2]) are in short supply — hence a dynamic addressing scheme, as employed by most of the dial IP service providers, is necessary;

- bandwidth contention must be employed in the network in order to cut down on network costs and thereby make a viable business case for offering broadband IP services;

- for some sectors of the market it is desirable to be able to offer an on-the-fly service provider selection capability (a possible exception to this is in a wholesale commercial model — see section 7.5.1).

Taking into account these and other factors, BT's broadband IP platform has evolved, during its trial stages, to a network whose architecture is shown at high level in Fig 7.9.

---

[2] RIPE, Réseaux IP Européens, is the body that administers public IP address allocations in Europe.

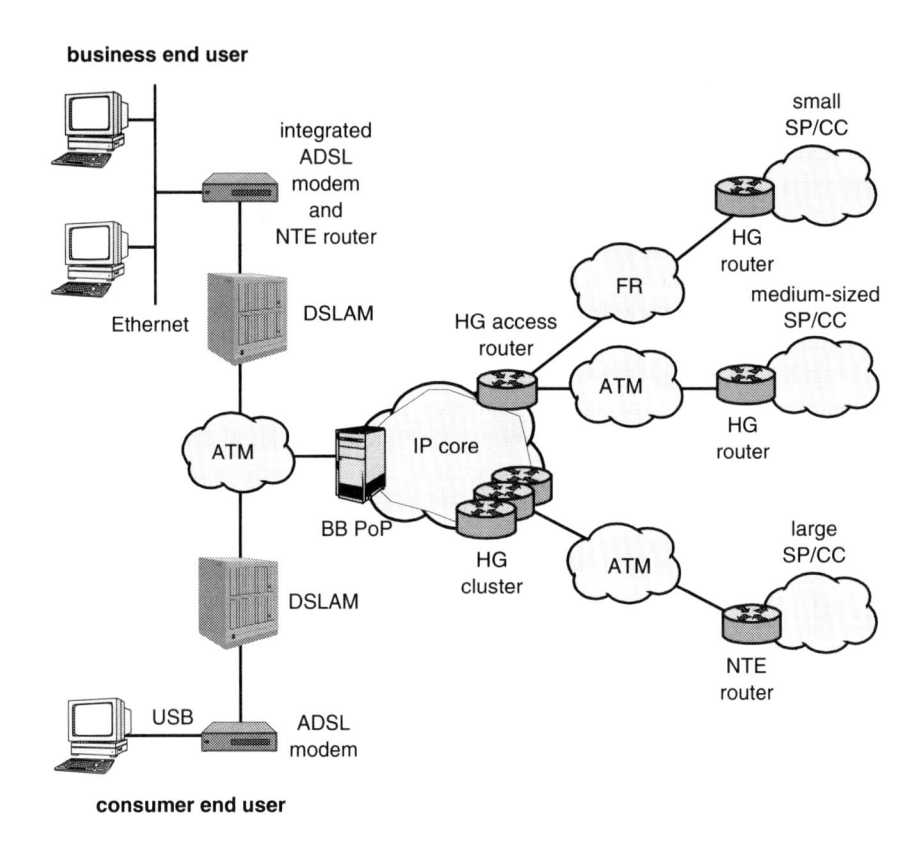

**Fig 7.9** BT's broadband IP platform.

### 7.5.1 Wholesale Broadband IP Products

Currently BT operates a wholesale commercial model with respect to all its broadband services. In this model a customer wishing to take a broadband IP service purchases it from a broadband IP service provider rather than directly from BT. (Of course, this does not preclude other parts of the BT group from retailing broadband IP services such as that done by BTopenworld. However, it is then important that the part of BT that operates the broadband network platform treats BT and non-BT broadband service providers alike in a fair and equitable manner.) It is then up to the service providers to purchase from BT the wholesale broadband IP products necessary to deliver service to their customers (hereafter referred to as 'end users'). In addition to commercial service providers, corporate clients (CCs) can make use of the wholesale products to provide their teleworkers with fast remote access to their internal network (intranet).

Two broadband IP products are required in order to be able to offer service to an end user. The first provides an individual access 'pipe' (based on an ATM permanent virtual circuit or PVC) from an end user's premises to their nearest broadband IP point of presence (PoP). The product has a built-in bandwidth contention to take advantage of the bursty nature of IP traffic and the fact that not all end users will be active at the same time — thereby allowing a significant reduction in network infrastructure costs. An instance of this product is purchased from BT by an SP (or CC) on behalf of and for each of their (end-user) customers. There are a number of options with regard to access speeds, network contention and the interface used for end-user presentation. The business quality variants, aimed at the SME market, are presented on 10baseT Ethernet and there are currently three speed options — 500/250 kbit/s, 1 Mbit/s/ 250 kbit/s, and 2 Mbit/s/250 kbit/s. An entry level variant, aimed at the (singleton-user) consumer market, will also be available. This will be presented on the universal serial bus (USB) and initially there will be only one speed option (500/250 kbit/s). In each case the quoted bit rates refer to the (downstream/upstream) IP throughputs rather than the ATM throughputs which are approximately 10% higher.

The second product that is required is an aggregate 'fat pipe', called BT Central, that provides connectivity between the broadband IP platform and an SP's premises. One or more instances of this product are purchased by the SP to carry the aggregated traffic from a number of their customers. These products are more like existing leased line type products — they are symmetric, with speeds ranging from 512 kbit/s up to 155 Mbit/s and various resilience options. As can be seen in Fig 7.9, there are two options for the underlying transport used to deliver these products to the customer (SP/CC). Frame relay (FR) is used to deliver the lower speed products (up to 2 Mbit/s) for which it is more efficient than the ATM used to deliver the higher speed products. The number and speeds of such products that an SP chooses to take depends on factors such as:

- how many concurrent end users they wish to support;

- whether or not they wish to impose further bandwidth contention on their services (over and above those already imposed by the end user access products);

- whether or not they wish to offer their customers different levels of service (say three offerings labelled as 'bronze', 'silver' and 'gold' representing three differently contended services).

### 7.5.2  PPP Sessions Over a Broadband Network

In order to resolve some of the problems highlighted at the beginning of this section (in particular those related to IP addressing and service selection), a session-oriented approach is taken to the provision of broadband IP services — much as is used in the world of dial IP services. The point-to-point protocol (PPP) is used to establish end-to-end sessions between an end user and an SP over which IP traffic can be transported. As in the case of dial IP, end users' IP addresses can be dynamically allocated during the establishment of a PPP session from an address pool administered by the SP. The network operator takes no part in this process thereby allowing them to avoid the significant overhead that would otherwise be associated with obtaining and administering vast allocations of IP addresses. In addition, the SP can reclaim IP addresses on termination of a PPP session (which can be end-user initiated or as the result of an idle time out), thereby allowing it to make efficient use of public IP address allocations.

The use of PPP sessions means that, strictly speaking, the resulting broadband IP services are not 'always on'. However, the time taken to establish a PPP session over a broadband network is typically very short (usually no more than a couple of seconds). In addition, provided that the underlying network has been dimensioned in such a way that there is always a usable minimum set of (albeit contended) resources available to every end user, then they should never, under normal conditions, encounter the broadband equivalent of a 'busy tone'. In this case the service is said to be 'always available'. A key feature of the architecture is the broadband PoP, the key component of which is one or more broadband access servers (BASs). These act as a brokering point with regard to the establishment of PPP sessions allowing end-user traffic to be directed to their SP on a per-session basis. From the PoP, IP tunnels[3] are used to provide point-to-point connectivity across a high-capacity core IP network to home gateway (HG) routers situated at or near the premises of the various SPs.

The PPP sessions originating from end users are extended via the appropriate tunnel to the SP and terminated (following successful authentication) on their home gateway(s). It is these tunnels along with appropriately dimensioned and provisioned access pipes and the HG routers themselves that constitute the aggregate fat pipes described above. The end-to-end architecture and associated protocol layering is shown in Fig 7.10.

There are a number of approaches to implementing PPP functionality on broadband networks. The approach preferred by the DSL Forum and that taken by BT is based on the PPP over ATM (PPPoA) recommendation (RFC 2364) drawn up by the IETF.

Although PPPoA is well established and many vendors have implemented it, it nevertheless has a significant disadvantage in so much as it does not support the multiplexing of a number of concurrent PPP sessions into a single ATM VC. This means that in an ATM environment which only supports PVCs it may not be possible to rapidly configure a new concurrent instance of an IP service given that the lead time to configure a new end-user PVC over the outer core ATM network (via management systems) may be of the order of days.

---

[3] The IP tunnels are implemented using the IETF recommendation for the layer-2 tunnelling protocol (L2TP) (RFC 2661).

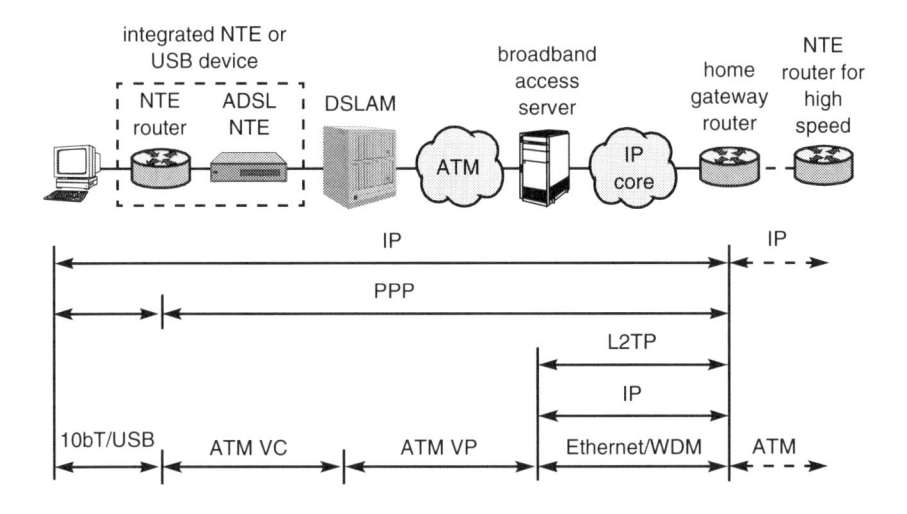

*Delivery of IP Over Broadband Access Technologies* 111

**Fig 7.10** End-to-end broadband IP architecture.

One way to overcome this limitation is to operate PPPoA in an ATM environment that supports SVCs. The full PPPoA recommendation supports the signalling protocols necessary to set up an ATM SVC during the establishment of a PPP session. SVC functionality in DSLAMs and outer core ATM infrastructure is still in development and may not be deployed for another two years.

There is another approach which is currently used by many US DSL network operators. This is to make use of the more recent IETF recommendation for PPP over Ethernet (PPPoE (RFC 2516)) which, as its name suggests, facilitates PPP sessions over a (connectionless, broadcast) Ethernet LAN. What is more, it allows the establishment of multiple concurrent PPP sessions and the auto-discovery of broadband access servers that support PPPoE. For this solution to work in the architecture shown in Fig 7.10, the NTE router would have to function as a transparent bridge with the bridged (layer-2) Ethernet frames then being transported over a single ATM VC to the broadband access server using a suitable layer-2 encapsulation such as the IETF recommendation for Multiprotocol Encapsulation over ATM Adaption Layer 5 (RFC 1483).

There are, however, some security issues with this approach. If not configured correctly, the BAS could bridge end-user connections so that layer-2 broadcast frames (such as those used by the address resolution protocol (ARP)) from a given end user might be broadcast to all the other end users served by that particular BAS. Apart from the associated security implications, this would also lead to a significant amount of unnecessary traffic being transported over each of the end-user VCs, thereby leading to significantly degraded performance of their IP service(s).

### 7.5.3 Network Address Translation

As shown in Fig 7.9, the service presentation to business end users is on 10baseT Ethernet via an NTE router. This allows a small customer LAN to be used to share the broadband IP connection among a number of hosts. In a PPPoA environment the PPP session is terminated in the NTE router. A simple Web server built into the NTE router facilitates easy configuration (of the username and password parameters) and control of the PPP session. The IP address obtained from the SP address pool during session negotiation is assigned to the network-side (or WAN) port.

There are then two options with regard to the IP addressing used on the LAN side of the NTE router. One option is to use private addressing (for example, as described in RFC 1918) and use network address translation (NAT) to hide the entire private IP subnet behind the single network-side IP address. The advantages of using NAT are that it makes efficient use of an SP's address allocation and offers the end user a high level of security against hacking. However, NAT is a double-edged sword — it does not allow unsolicited IP access from the network-side of the NTE router. This is problematic for an end user taking an Internet access service who wants to host Internet services (Web servers, etc) on their own premises.

There is an expedient alternative that can be used to address this issue. An approach variously known as 'static PAT (port address translation)' or 'punctured NAT' can be used to define specific UDP/TCP port numbers that will be 'listened for' on the network-side interface. These are mapped to specific IP addresses on the LAN-side subnet which correspond to hosts upon which the required Internet servers are running. The problem here is that as each new Internet service is added by an end user, it is necessary for the network operator to make a change to the configuration of the NTE router.

Another aspect to NAT which makes it undesirable for many end users is that it is not transparent to all higher layer applications. In particular, H.323-based videoconferencing applications and many network games will not work.

The other option for addressing used on the LAN is to avoid running NAT on the NTE router — thereby eliminating the problem of application non-transparency described above. In order for this to work, an IP subnet, which is part of the SP's address allocation, has to be statically assigned to the end-user LAN and a suitable IP route established on the HG router. Of course, this approach means that the IP addresses spanned by that subnet are unavailable for use with other end users' PPP sessions. This option is therefore expensive with regard to IP address usage. In addition, the NTE router no longer has the security that comes with NAT (unless firewall functionality is put in its place).

At the launch of BT's broadband IP products there will be both NAT and 'no-NAT' options for the product variants aimed at the business end-user market.

The preceding arguments are not applicable to consumer end users. This is because the entry level broadband IP product uses the USB to extend the ATM VC from the ADSL modem into the host itself (in essence the USB ADSL modem is acting as a layer-2 bridge). This allows the PPP session to run all the way from the SP's home gateway router to the end user's host. The IP address allocated by the SP during session negotiation is assigned directly to the host as in dial IP. (In fact the method of managing the PPP session is very much like the familiar 'dial-up networking' client.)

It is possible for a consumer end user to share their single IP connection with additional local hosts (via Ethernet, for example) by making use of readily available router software that will run on their directly connected host. However, this software will necessarily have to include NAT functionality.

### 7.5.4 Evolution of the Broadband IP Platform

As the QoS and security features of IP are enhanced and routers become ever faster it is not unreasonable to envisage the broadband IP platform evolving towards a fully routed IP network[4] extending out to the DSLAMs (and non-wireline equivalents) and even as far as an end user's CPE. Indeed it makes a lot of sense to put IP awareness into DSLAMs (especially multicast capabilities), since their locations are ideal replication points for the distribution of streamed multicast video. (In the event of IP video streaming services really taking off — as many now predict — this approach would offer significant reductions on the need for expensive core capacity that would otherwise be used up by multiple unicast IP video streams destined for end users served by the same DSLAM.)

The enhancements in security and the massive increase in address space that will come with the introduction of the next generation IP (IPv6) will also have a significant effect. This will effectively remove the requirements for session-based end-user access and the use of IP tunnels in the core (thereby removing the protocol and processing overheads that come with them). In addition, protocols such as the IETF resource reservation protocol (RSVP (RFC 2750)) are introducing the methods necessary to be able to access the QoS capabilities (previously the sole domain of ATM) that are now appearing in IP switch-routers.

However, this vision is still some way off. Estimates as to when IPv6 will be extensively deployed vary from two to ten years from now. Either way, when deployment does begin in earnest there will no doubt be a problematic period of migration from IPv4.

---

[4] For the same reasons as have already been discussed in section 7.4, this may well be in conjunction with the introduction of MPLS as an expedient way to integrate IP and ATM.

## 7.6 The Evolution of Broadband Wireline Access Networks

The lowest common denominator for the evolution of access networks is that they should provide ever 'fatter pipes'. This simplistic view is correct up to a point, but the demands that such networks must have the capability to support many IP and other services with varying characteristics mean that broadband access networks will have to support increased functionality, configurable features and intelligence. The FSAN (full service access networks) initiative was set up in the mid-1990s with the vision of creating a world-wide common requirements specification for the next generation of fixed access networks [5]. Common themes to emerge from FSAN are the requirements for deeper fibre penetration and multiservice support.

### 7.6.1 Fibre in the Local Loop (FITL)

The FSAN approach to bringing fibre closer to the customer is based on a shared optical fibre feeder network called an APON (ATM passive optical network). An APON optical line termination (OLT) would feed a number of optical network terminations either on the customers' premises or at intermediate points in the access network. The main advantage of APONs over existing point-to-point optical fibre networks is that the interface electronics of an APON head-end are shared between many optical network terminations, thus reducing interface costs per customer. An APON network is broadcast in nature and requires a media access control (MAC) protocol to administer the ingress of upstream traffic on the PON. The FSAN requirements specification envisages the use of ATM to allow a range of traffic types, e.g. real-time, circuit-emulation, IP and quality-of-service levels. The perceived need to support different QoS guarantees and traffic types for many of the connected customers required layer-2 functionality. The challenge to FSAN network operators now is to support emerging end-to-end IP services.

The emergence of IP as the dominant service-carrying protocol has ramifications on the use of FSAN-type FITL networks. The bursty nature of much upstream IP traffic combined with the shared upstream access nature of the PON means that some existing 'static' MAC protocols (Fig 7.11) may inhibit the flow of upstream IP traffic on a contended or overbooked PON. The use of 'dynamic' MAC protocols may help to alleviate such problems on a highly contended PON [6]. The need to support more IP traffic, which is much more symmetric in nature than the highly asymmetric video-type applications, may also drive a requirement for PONs with larger upstream capacity. Studies have shown a modest component cost increase per ONU for a 622 Mbit/s symmetric PON.

A longer term approach may be to introduce wavelength division multiplexing (WDM) on a PON to provide additional capacity without having to build new fibre. Wavelengths could then be allocated to a particular ONU which would be equipped with a tunable receiver. A further likely development will be the integration of IP

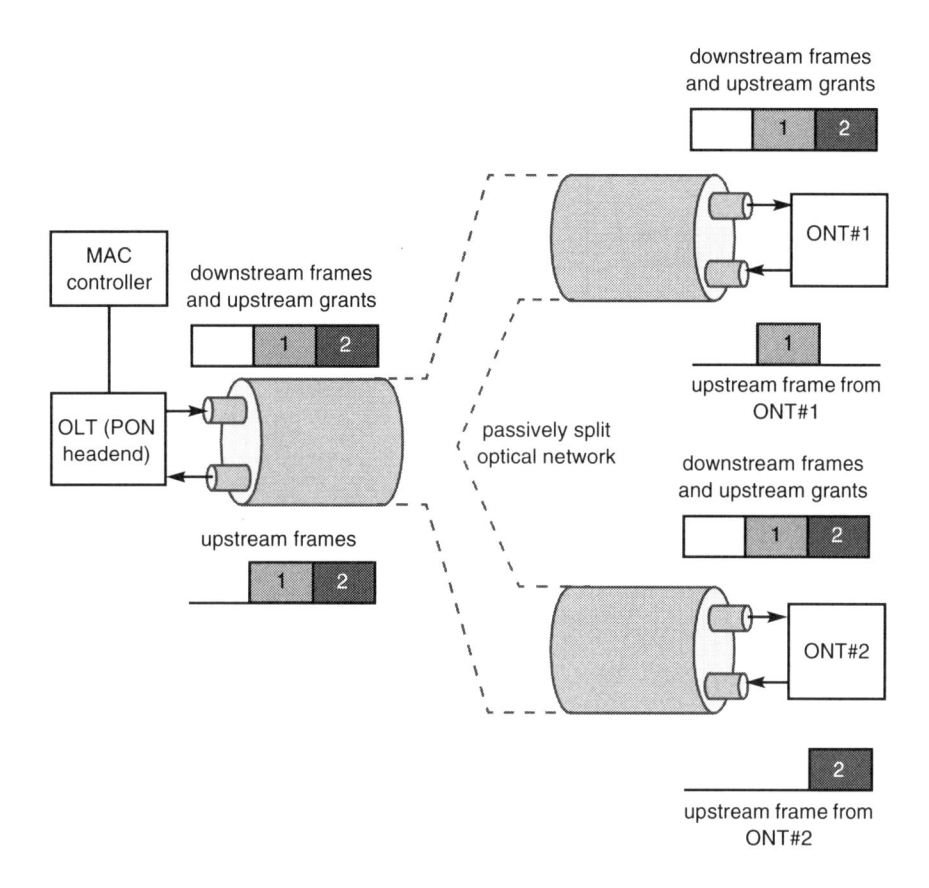

**Fig 7.11** The use of media access control on a PON to administer upstream access.

functionality, such as routing and IP QoS into the OLT and the ONU/T. The development of an IP-PON would require QoS issues to be resolved for all services and new MAC protocols to be developed. An optical overlay network based on WDM could provide a way to achieve an IP-PON type functionality upgrade.

### 7.6.2 VDSL

Fibre in the loop architectures may either connect fibre directly to customers via an optical network termination (ONT) or to an intermediate point in the network, thus 'shortening' the distance to the customer. The FSAN specification envisages the use of the 'fibre to the 'x' ' architecture where 'x' is an active node in the access network (Fig 7.12). Very high speed digital subscriber line (VDSL) would then be used over the final copper drop to the customer. VDSL is fundamentally an

116   *The Evolution of Broadband Wireline Access Networks*

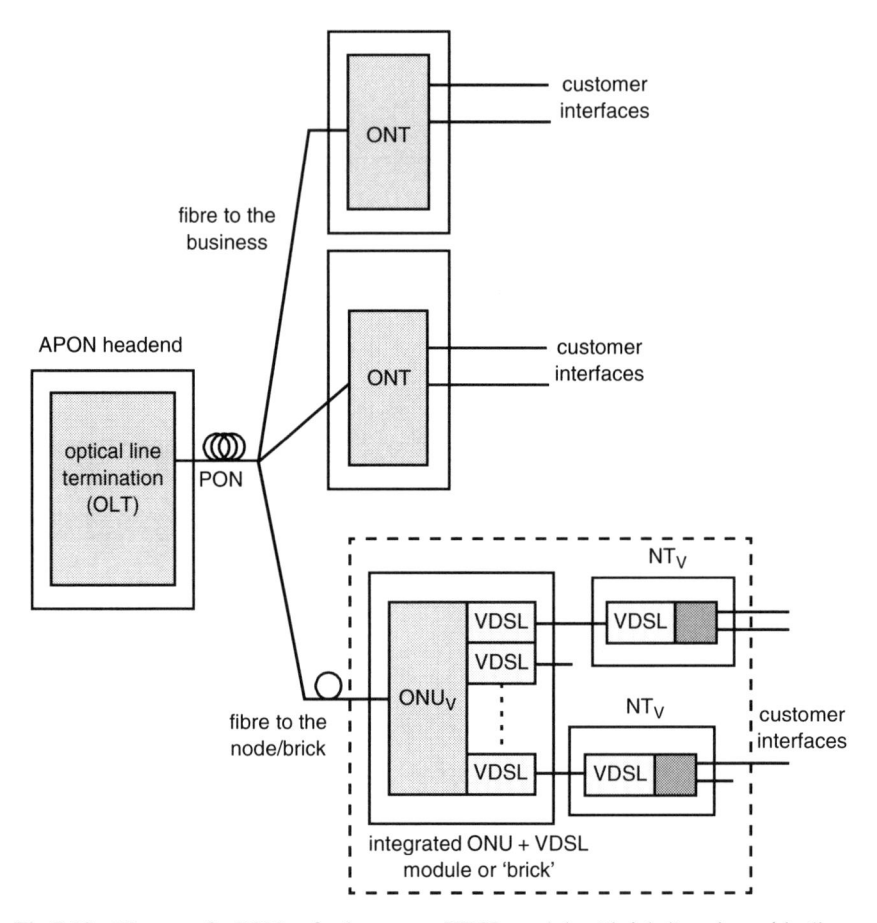

**Fig 7.12**   The use of a PON to feed compact VDSL modules ('bricks') and provide direct fibre connections to customers.

extension of ADSL modem technology enabling higher (than ADSL) data rates over copper telephone lines. The main differences between ADSL and VDSL are that the latter uses wider band transmissions at lower power levels, can be operated in symmetric mode, and has a shorter effective range over copper telephone lines. However, both flavours of DSL can potentially be operated in the same cable set.

The large upstream capacity and the use of VDSL in symmetric mode would be particularly suitable to provide IP services that have symmetric bandwidth needs. This would require the bundling of multiple services over VDSL, such as multiple voice lines (VToA or VoIP), Internet access and Web hosting. The requirement to provide a circuit-type functionality for voice, for example, may facilitate the early introduction of IP functionality into an integrated VDSL NT.

### 7.6.3 Other Trends

The use of wireless interfaces on customer equipment and the next generation of mobile technology will have a profound impact on the functionality of emerging access networks. The need for seamless handover between the fixed and mobile networks, common applications, and multiple concurrent access per terminal for IP services, all have significant ramifications on the architecture of the broadband access platform. With the increasing bandwidth available from broadband wireline networks based on APON and VDSL, several services will be bundled. This means that any solution to managing the network and retaining QoS must be scalable.

The low customer density for many rural locations implies that investments in either DSL or optical fibre infrastructure may never earn a return. Investigations into alternative techniques centre around fixed wireless (for which sufficient radio spectrum is not always available) or the use of satellite delivery with a low-speed return channel provided via the PSTN or ISDN.

Incumbent telco and cable company access networks have both begun to have an increasing amount of optical fibre infrastructure over the last few years. This enables homes and businesses to have improved broadband connectivity for future communications needs.

However, for years the 'bread-and-butter' business of telcos has been delivery of POTS using an access network that is largely copper based. This then evolved first to the use of ISDN and then ADSL to deliver increasingly broadband digital services over the existing copper access infrastructure. Investigations have shown that a fibre-to-the-cabinet architecture could potentially be a cost-effective broadband delivery solution [7] which further improves the access network capability. This high-speed access architecture could play a significant role in bringing multiple channels and higher quality channels, such as for HDTV, to those customers who require them. This will be required in the future as interactive multimedia services proliferate. Each member of a family may wish to access such services simultaneously, e.g. one member watching a movie, another shopping from home and children playing games or getting homework assistance over the network.

### 7.7 Conclusions

There are a number of broadband access technologies which are capable of effectively providing access to broadband IP services. These technologies and their associated network platforms will continue to evolve. Broadband access is no longer just a 'fat pipe' — many systems have increased functionality, configurable features and intelligence. Hence issues such as service provisioning, interoperability and CPE auto-configuration are increasingly important challenges that need to be overcome for mass-market viability.

The various broadband access technologies have their relative merits and all will continue to prosper as the insatiable customer demand for bandwidth continues. Many operators and service providers will use a combination of these technologies in order to best approach their target markets. In the UK some cable operators already use DSL on their 'siamese' cables which include coaxial cable (used by cable modems) and twisted copper pairs.

ISPs are using both DSL and broadband radio in the USA to increase customer coverage. Radio systems in conjunction with in-building distribution via DSL are being used to target office blocks. Satellite is increasingly being examined to expand broadband coverage to rural areas.

BT's current ADSL platform is well positioned to take advantage of the proliferation of broadband IP services. The platform has the capability to increase its functionality and evolve towards incorporation of APON and VDSL wireline technology.

## References

1  Foster, K. T. et al: '*Realising the potential of access networks using DSL*', BT Technol J, **16**(4), pp 34-47 (October 1998).

2  Jewell, S., Patmore, J., Stalley, K. and Mudhur. R,: '*Cable TV technology for local access*', BT Technol J, **16**(4), pp 80-91 (October 1998).

3  Merrett, R. P., Beastall, P. V. E. and Buttery, S. J.: '*Wireless local loop*', BT Technol J, **16**(4), pp 101-111 (October 1998).

4  Williamson, J.: '*High hopes*', Global Telephony (January 2000).

5  Quayle, J. A., et al: '*Achieving global consensus on the broadband access network — the full service access network initiative*', BT Technol J, **16**(4), pp 58-70 (October 1998).

6  Hoebeke, R., Ploumen, F. and Venken, K.: '*Performance improvements in ATM PONs in multi-service environments by means of dynamic MAC protocols*', Broadband Access & Network Management, NOC'98 (June 1998).

7  Olshansky, R. and Veeneman, D.: '*Broadband ADSL*', ADSL Forum contribution ADSLForum95-016 (March 1995).

# 8

# WIRELESS ACCESS

## C Fenton, B Nigeon, B Willis and J Harris

## 8.1 Introduction and Scene Setting

This chapter describes wireless technologies that can be used for carrying IP data. UMTS is one of these technologies; it has had a high profile due to the cost of its licences but is also another technology that may change the shape of the Internet with its ability to bring IP applications to potentially billions of mobile terminals. Other wireless technologies with a potential to change our lives are wireless LAN and Bluetooth — wireless LAN may free many users from plumbing their offices and homes with cables, while Bluetooth will enable a new type of data network, a data network for our personal space.

The chapter gives the reader a useful insight into these upcoming wireless technologies, several of which exist depending upon the application and the nature of the telecommunications carrier implementing them. On the one hand, wireless systems can be used to provide mobility, getting rid of the link with a fixed location. On the other hand, wireless systems can be used by a telecommunications carrier as part of its infrastructure when it is more convenient or cheaper to set up a wireless network instead of a wired one. The following three types of wireless system are investigated here:

- cellular systems which provide users with mobility based on a mobile-enabled network;
- cordless systems which provide in-building coverage as a short range wireless extension to a fixed network — with cordless systems, users enjoy local mobility while in the building;
- fixed wireless access systems which provide a telecommunications operator with a last mile wireless drop between its point of presence on a fixed network and the customer — the 'wirelessness' is only in the network infrastructure and is not apparent to the end-customer.

Data services on these wireless systems have developed rapidly over the last five years. They have changed due to many drivers and enablers, mainly technical, economical and regulatory. The chapter discusses these enablers and highlights what is planned and when it might be available.

## 8.2 Wide Area Public Cellular Developments

Since the 1992 WARC decision to allow the operation of 3rd generation mobile radio in the 1900—2200 MHz band, operators, manufacturers and national administrations have been working to develop suitable technologies. The ITU set up IMT2000 (International Mobile Telecommunications for the year 2000), a programme to ensure harmonisation between proposed 3rd Generation systems. In December 1999 the IMT2000 agreed on the following technologies:

- IMT-DS — based on ETSI W-CDMA, using direct spread code division multiple access (CDMA);

- IMT-MC — based on US cdmaOne, using multi-carrier CDMA;

- IMT-TC — based on ETSI TD/CDMA, using time and code division multiple access;

- IMT-SC — based on UWC-136/EDGE (enhanced data rates for GSM and TDMA/136 evolution);

- IMT-FT — based on DECT.

There are different technologies to address the fact that there are paired and unpaired spectra, and to allow backward compatibility with the dominant 2nd generation systems (GSM and cdmaOne). The spectrum allocation is shown in Fig 8.1 (in section 8.2.4), where the paired and unpaired blocks can be seen, together with the satellite band. The band between 2025 MHz and 2110 MHz is already assigned to other services. The paired spectrum will use frequency division duplex (FDD), where one carrier is designated for uplink and one for downlink. The unpaired spectrum uses time division duplex (TDD), where different time-slots on the same carrier are used for up and downlink; IMT-TC and IMT-FT are designed for TDD operation, the others are for FDD.

Within Europe, operators can choose to use any of these technologies for their 3rd generation system, with the caveat that at least one operator in each country must use IMT-DS in order to provide roaming capability. Within the UK and Europe, the most likely choices will be IMT-DS and IMT-TC. Operators with existing cdmaOne networks are likely to choose IMT-MC, which includes much of North and South America. However, in the USA the spectrum designated by WARC '92 is currently in use by 2nd generation systems, so the vision of a single global terminal is still some way off. Japan and much of Asia/Pacific is expected to use IMT-DS.

### 8.2.1 Achieving Higher Bit Rates

GSM is a narrowband system capable of supporting a maximum data rate of 9600 baud for a data call, which to date has been based on a circuit-switched architecture, with a core network based upon ISDN with 64 kbit/s switching. The radio interface is a hybrid frequency and time division multiple access (F/TDMA) system, using a 200 kHz carrier with Gaussian minimum shift key (GMSK) modulation. Each carrier is divided into eight time-slots, and it is one time-slot that is used per voice or data call. In order to achieve higher data rates, it is necessary to at least modify the way in which the radio interface is used, and at most implement a completely new air interface and modulation scheme. This requires new base-station equipment and new terminals.

### 8.2.2 UTRA W-CDMA

The UTRA technologies are completely new radio systems, based on CDMA with quadrature phase shift key (QPSK) modulation. In CDMA systems, rather than sharing the radio resources in the frequency or time domains, it is shared in the power domain. The IMT-DS system, more commonly known as W-CDMA, uses a 5-MHz carrier (compared to 200 kHz for GSM). The absolute maximum data rate that a single carrier can support is 2 Mbit/s, but this is limited by the practicalities of implementation and the hostile radio environment. In reality, the maximum circuit-switched data rates that are likely to be achieved are in the order of 144 kbit/s.

However, the 3rd generation systems will also support packet-switched operation, based on the general packet radio service (GPRS). By taking advantage of the statistical multiplexing that is inherent in packet-switched systems, data rates of the order of 384 kbit/s are likely to be achieved. The W-CDMA system is designed for wide-area coverage, although the cell sizes will be significantly smaller than those for GSM, due to the higher frequency and the physical limits of the radio interface supporting high data rates. Given that the cells will be smaller, it will require a significant increase in network infrastructure to provide equivalent national coverage to existing GSM networks. However, it will be possible to handover calls between GSM and W-CDMA, although with reduced capabilities. This will allow operators to offer national coverage through a single dual-mode terminal, without having to deploy massive numbers of base stations.

### 8.2.3 UTRA TD/CDMA

The TD/CDMA system complements W/CDMA, in that it is designed for use in the unpaired spectrum. The advantage of TD/CDMA is that it supports high bandwidth asymmetric traffic much more efficiently than W-CDMA. In TD/CDMA it is possible to dynamically assign time-slots to be uplink or downlink; therefore, if there is a greater demand for downlink capacity than uplink, it is possible to support this by assigning more time-slots to the downlink. This is not possible in W-CDMA, where an uplink and downlink carrier is always needed. This is particularly relevant to applications such as e-mail/voicemail retrieval and Web browsing. However, due to the spectrum allocation there are interference issues associated with TD/CDMA that restrict the deployment flexibility, and it is likely to be used for indoor installations providing high bandwidth pico-cellular coverage to offices, public buildings and shopping centres, for example.

Unfortunately, the development of TD/CDMA is lagging behind that of W-CDMA. This is because the manufacturers are concentrating their efforts on the Japanese market, where there is an urgent need for W-CDMA equipment to relieve the capacity problems they currently have with their 2nd generation networks. Compounding the issue is the fact that there is no TDD spectrum available in Japan, hence no requirement for TD/CDMA equipment.

### 8.2.4 The GSM Future?

What of GSM apart from providing a fall-back solution where there is no 3rd generation coverage? There are two developments that build on the existing GSM system that will allow data rates in excess of 100 kbit/s from a GSM carrier — high-speed circuit-switched data (HSCSD) and GPRS. Rather than change the fundamental radio interface, these services change the way the radio resources are used by the network. HSCSD uses multiple time-slots in a circuit-switched configuration, whereas GPRS uses multiple time-slots and can multiplex users on the same time-slot. The other fundamental difference is that being packet switched, GPRS allows the user to operate in an 'always on' session. HSCSD users will tend to log on and off as required, in order to avoid a massive bill. Both of these systems offer flexible support for asymmetric traffic load, and require no change to the radio equipment in the network. Of course new terminals will be required, together with additional equipment in the network to support the packet switching and higher rate circuit switching. BTCellnet is planning to launch GPRS in its network in the summer of 2000. GPRS and HSCSD will have a role where 3rd Generation will not be launched for some time — or by operators who do not acquire 3rd Generation spectrum. GPRS is preferred over HSCSD since, being packet based, it makes more efficient use of the spectrum. Few if any operators have plans to deploy HSCSD.

The major development for GSM is EDGE (enhanced data rates for GSM evolution). This is an entirely new air interface, but still within the 200 kHz channel bandwidth of GSM. EDGE has a new modulation scheme (8PSK) which can transmit 3 bits per symbol rather than the one of GMSK, and nine coding schemes, supporting data rates between 9 kbit/s and 60 kbit/s per time-slot. EDGE also supports time-slot concatenation, thus raising the maximum data rate to 473 kbit/s when all eight time-slots are used. EDGE is only the air interface, and as such can support HSCSD or GPRS. Again this technology is key for operators who do not acquire 3rd Generation spectrum and need to compete as best they can. However, there will be a significant cost to develop a network for EDGE, as it will require new equipment in existing base stations, and a large number of additional base stations, which is the major cost of the cellular network infrastructure.

### 8.2.4.1 The UK spectrum auction

The development of 3rd generation systems is being driven, not by the operators, but more by the governments who are keen to ensure they obtain the maximum return from this valuable resource. In order to stimulate further competition, the government has decided to ensure that there is at least one new mobile operator in the UK. This has determined the way in which the available spectrum had been packaged for the licence auction (the spectrum packages are shown in Fig 8.1). This shows that the most attractive package is reserved for a new entrant, having three FDD carriers and one TDD carrier.

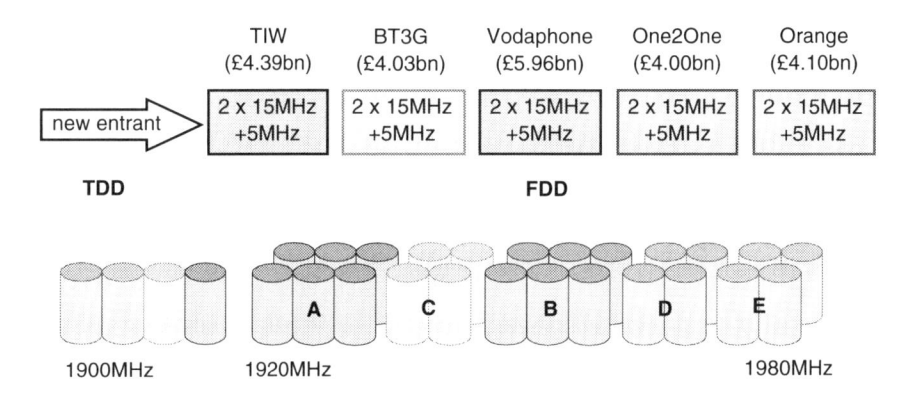

| | TIW (£4.39bn) | BT3G (£4.03bn) | Vodaphone (£5.96bn) | One2One (£4.00bn) | Orange (£4.10bn) |
|---|---|---|---|---|---|
| new entrant | 2 x 15MHz +5MHz | 2 x 15MHz +5MHz | 2 x 15MHz +5MHz | 2 x 15MHz +5MHz | 2 x 15MHz +5MHz |

**TDD**                                      **FDD**

1900MHz     1920MHz    A    C    B    D    E     1980MHz

**Fig 8.1** The spectrum allocation.

This is designed to offset the advantage that the existing operators have with their existing spectrum and networks. The next most attractive package is the one with three FDD carriers. This will allow the flexible deployment of a high capacity network. This was borne out by Vodafone's willingness to pay such a high premium for this licence. The remaining three packages of two FDD carriers plus a TDD carrier do not offer the same flexibility as the other packages, and are likely to require a greater investment in order to meet the capacity requirements. In the UK, it has been the government's intention that we should be one of the first countries to launch 3rd generation systems, and so a target launch date of January 2002 has been specified. With the licences awarded in late April 2000, and the standards still under development, the risks to the operator are significant. The effort required to deploy a viable network in the time-scales will be significant, and there is little or no experience in the design and deployment of this technology. So while being first to market is essential in terms of the business model, the technical and practical challenges will be significant, and being first might not mean being the winner.

### 8.2.5 The Global Position

Given that UMTS was originally proposed as a global system, it is somewhat disappointing that there are compatibility issues with both the technology and the spectrum. The main issue is that in the US there are 2nd generation mobile networks using part of the UMTS band. This will perpetuate the situation of today whereby a multi-band terminal is required for operation between the US and most of the rest of the world. While this has been solved in recent advances in handset technology for GSM, operating at 900 MHz and 1900 MHz, the problem is not so simple for UMTS — European terminals will have to support IMT-DS, perhaps IMT-TC, and GSM900/1800. In addition, a global UMTS terminal will have to support IMT-MC and possibly GSM1900.

However, the story does not end there. There are operators, especially in the USA, that have accepted the fact that they will not get access to UMTS spectrum, and as such are planning to deploy EDGE in their existing spectrum, at 800 MHz. This situation will not be limited to the USA. Many existing GSM1800 operators are likely to deploy EDGE in their existing spectrum, since, in the absence of UMTS spectrum, EDGE allows an operator to offer data rates approaching those of UMTS. This now adds an additional technology, and possibly RF band, for a global terminal to support.

In terms of the licence availability, much of the global activity is planned for 2000 and early 2001. The Japanese and Finnish spectrum was allocated in 1999. Across Europe and the Asia/Pacific region there are over 20 licences planned for sale before the first networks become operational in 2002.

BT's wholly owned subsidiary Manx Telecom, on the Isle of Man, was allocated a licence to operate UMTS services during 1999 and is planning on launching services in 2001. The key focus here is to enable the Isle of Man to showcase the capabilities and opportunities offered by 3G and for other parts of BT to link their capabilities through the UMTS access network.

### 8.2.6 GPRS in More Detail

General packet radio service is a development that changes the face of cellular access completely. To date GSM, which is the leading global cellular technology, has really only created a mature cellular access system. It has a comprehensive set of features from roaming to terminal independence, data services and full ISDN compatibility. However, it lacks basic channel bandwidth, and optimal connectivity for the data user. Essentially GSM is an ISDN system of limited bandwidth with a radio link between the CPE and the local exchange where the selection of the route to the local exchange is managed automatically. Key features are:

- ISDN service:

    — voice;

    — data (limited maximum data rate);

- roaming (automated attachment to friendly networks);
- use of the SIM (subscriber identity module).

GPRS is a development within the ETSI GSM community to work hand in hand with the increasing bandwidth capabilities that provides services that are directly in line with the IP data wave. Key features are:

- efficient use of radio spectrum:

    — bandwidth upon demand;

    — IP 'on-line' — always on and available for data receipt (but not always consuming radio resource);

- provision of access to the Internet — both public and private;
- higher data rates.

To deliver this, GPRS revolves around Internet protocol technologies and has meant a very significant change to the GSM architecture.

Figure 8.2 shows the essential key components. Data transfer is based upon a two-hop process one across the GPRS backbone network between SGSN[1] and GGSN[2], and the other between the SGSN and the mobile terminal. Mobility is managed on two levels.

- The route between the mobile and the SGSN

    This is updated regularly as the mobile moves or as capacity/congestion requires. The route is not maintained as for ordinary GSM circuit-switched calls, where for voice a conference bridge is used to reduce the detectability of the cellular 'handover'.

    For GPRS a route is 'reselected' and packet/data transmission is paused while the change of route is made (changes to TDMA time-slot and frequency).

---

[1] SGSN = serving GPRS service node.
[2] GGSN = gateway GPRS service node.

126  *Wide Area Public Cellular Developments*

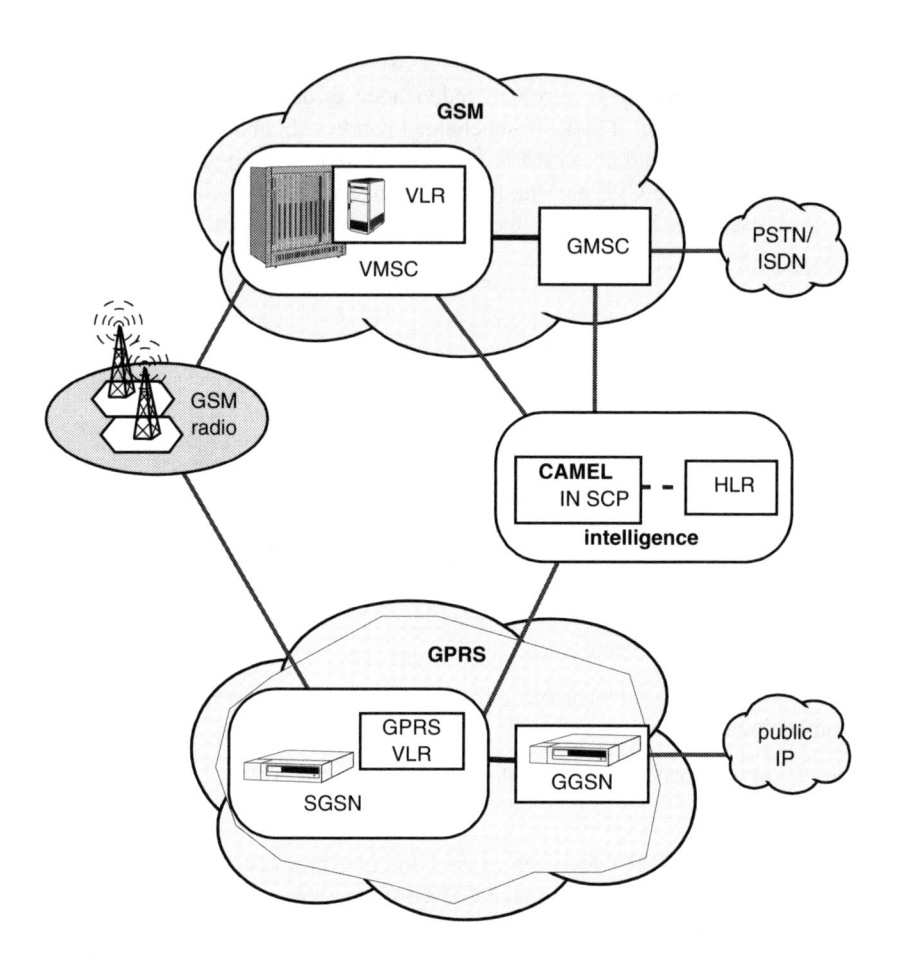

**Fig 8.2**  GPRS on GSM architecture.

- The route between the SGSN and the GGSN

  This is nothing more than an IP tunnel through a standard IP network (GPRS backbone) and is used to route packets bound for the mobile or to an Internet terminal quickly. If a mobile should move to a radio cell that is attached to a different SGSN, then a process starts to re-establish the tunnel on the new SGSN.

  Care is taken to re-route IP packets that might be in transit when the mobile reselects on a new SGSN and these are forwarded by the old SGSN to the new one through the GPRS backbone.

Much of GPRS is built upon existing IP technology and principles. Developing QoS and capacity is now where the development for GPRS must go and again many of the existing solutions are being considered.

A key issue for GPRS is that it is not going to be free. To date mobile is still able to charge a premium and the 'on-line' access that GPRS gives users to the IP community is going to be of high value. However, that does not mean that people will be prepared to pay heavily for it. Users' expectations are (falsely) that Internet is free. This will push mobile operators hard to change their business models in order to create the same perception (see iMode from NTT DoCoMo in Japan [1] where a new business model is working and commercially very successful).

Users will also be expected to pay for increased quality (both priority and data rate) which may be easier to justify and comprehend.

Table 8.1 depicts the basic GPRS bearer rates and it is important to bear a few factors in mind:

- the bearer rate is not the same as the user rate;
- deployment in the near term will allocate a few traffic channels per cell — this will increase as capacity builds up;
- radio channels will be shared between users;
- there is a limit of two traffic channels per mobile (i.e. mobile capabilities limited to communicating over 2 × traffic channel at any time — this is known as multi-slot operation).

Table 8.1    GPRS coding schemes and rates.

| Coding scheme | Encoded rate | Raw data rate (kbit/s) | Approx user rate kbit/s |
|---|---|---|---|
| **CS1** | 1/2 | 9.05 | 6 |
| **CS2** | ≈ 2/3 | 13.4 | 9 |
| **CS3** | ≈ 3/4 | 15.6 | 10 |
| **CS4** | 1 | 21.4 | 14 |

Note: 'Approx user rate' takes account of retry errors and signalling overhead — LLC/MM. For comparison, a current GSM data user gets 'something less than' a quoted rate of 9.6 kbit/s.

At the same time as GPRS deployment, cellular networks will be upgraded to support multi-slot. This allows the allocation of more than one traffic channel to a user for communications. Therefore the theoretical higher data rates quoted for GPRS become possible, i.e. 8 slots × 21.4 kbit/s = 160 kbit/s.

Overall GPRS is a key enabling technology that brings packet services, or the 'IP on-line' capability to the wireless community. Its existence will be key to the wider area systems and will grow as the networks are deployed and terminals are provided.

### 8.2.7 Migration to UMTS

The third generation system will develop from GSM and utilise the basic GPRS capabilities but with the new UMTS radio interface. Figure 8.3 illustrates the initial step to UMTS by evolving the GPRS enabled GSM network. The new UMTS radio interface is added and provides enhanced access capabilities through both the circuit-switched and packet-access routes.

This sort of migration also means that for an existing GSM operator they could launch a network with national coverage but with limited service capability in some areas (e.g. a lower data rate while on the GSM portion of the network). For some time the fact that operators have committed a large investment in the existing GSM networks will mean that they will wish to maintain the GSM system and customer base without further expense. For the capacity-hungry users, instant migration to the UMTS system will provide them with the sort of service they are looking for but without the national coverage. But for many of these people city-wide coverage would be enough.

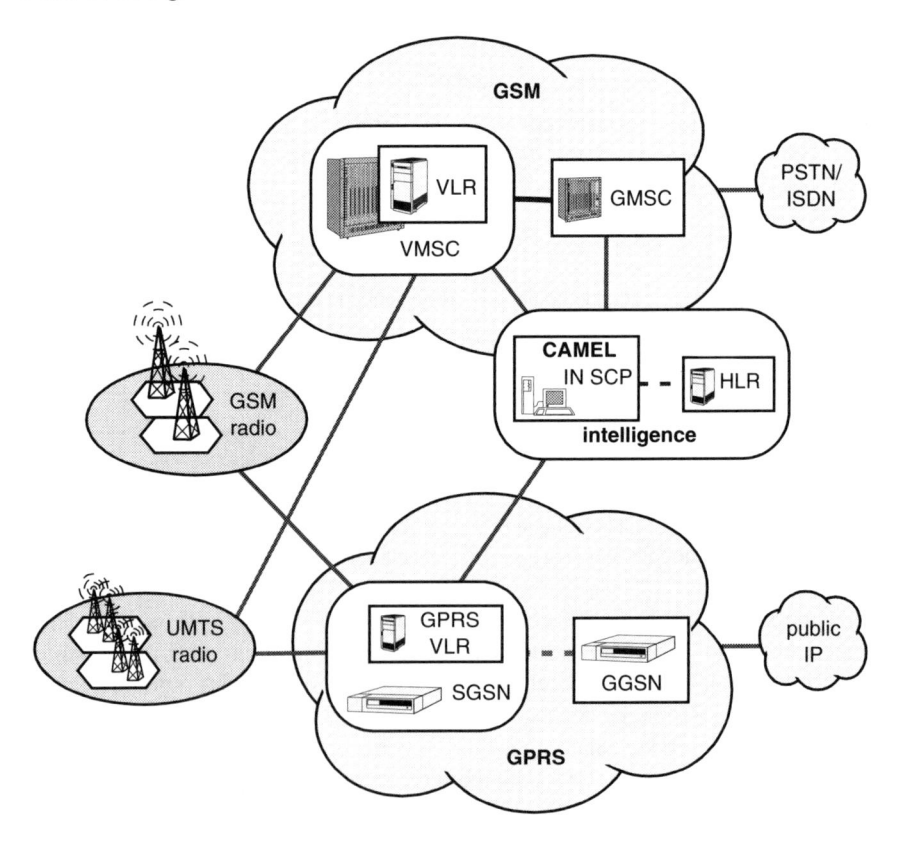

**Fig 8.3** UMTS evolving from GSM.

In the longer term the GSM spectrum will be reused and the UMTS network will eventually look like Fig 8.4.

This is an all-IP solution where direct interworking to the data networks is by far the majority service, but remaining legacy switched networks are supported through a media gateway for address and format translation.

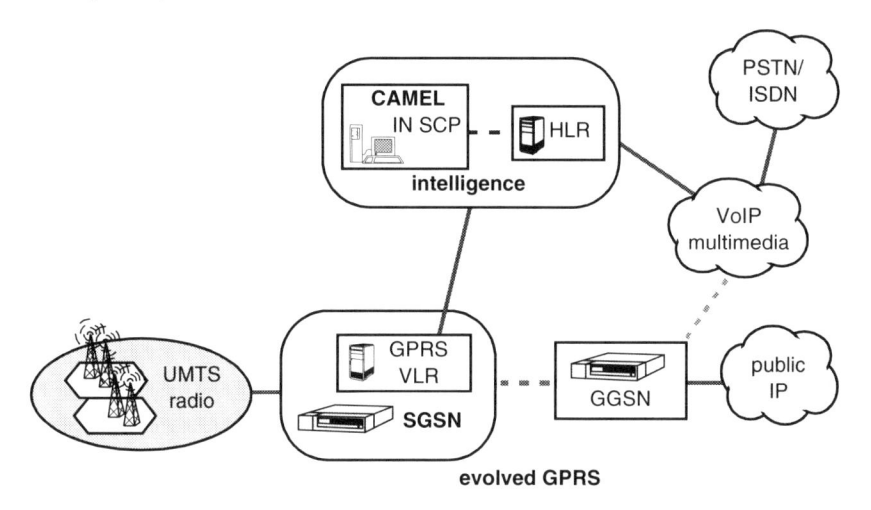

**Fig 8.4** UMTS all IP.

## 8.3 Cordless Access Developments

As users get accustomed to the mobility offered by cellular networks, they start to ask for the same flexibility at home and in their working environment. The massive success of DECT digital cordless telephones shows the customer's need for mobility for voice applications.

For data applications, the perception is that such flexibility is not achievable with current wireless data technologies — a perception endorsed by the small number of wireless data products that have been available on the market until recently. These were generally expensive and often not up to user expectations because of poor performance, lack of interoperability and poor market targeting.

Today, technologies have matured, standards have appeared and products are available that start addressing the above issues, making it possible to consider cordless technologies as an extension to a fixed network to provide local mobility for data applications.

The previous section presented the evolution of the cellular network to offer data services. This section presents the cordless technologies available today and their potential for integrating with IP networks to provide high-speed wireless data services.

### 8.3.1 Offering High Bit Rate Indoor Coverage

No cordless technology today can match the capabilities of the highest speed data networks. In fact, it is a fact of life that wired networks will always be ahead of wireless networks. It is not that it is theoretically impossible for wireless technologies to offer very high data rates (as proved by technology like LMDS), but that cordless products are usually owned by the end-customers, and the cost of this equipment needs to stay low in order to remain attractive. Therefore, trade-offs have to be made between capabilities and cost. Improvements in the capabilities of cordless systems are often due to more powerful technology becoming affordable rather than radical technological advances. Today's cordless technologies can offer bit rates up to 6 Mbit/s (IEEE802.11b Wireless LAN Standard). Future developments in the IEEE802.11 and HIPERLAN standards will soon make data rates up to 34 Mbit/s a reality. Most cordless technologies have been designed from the outset to interwork with packet data networks making the integration with wired networks very simple and seamless. Services available over the wired network will work identically on wireless networks with the benefit of mobility and the drawback of lower bit rate.

### 8.3.2 DECT

Digitally Enhanced Cordless Telecommunications (DECT) is a European standard which is now recognised and used world-wide. The standard provides a general radio access technology for wireless telecommunications, operating in the licence-exempt 1880 to 1900 MHz frequency band (in Europe). DECT is mainly used today for voice (cordless telephones) and wireless local-loop systems (see section 8.4). DECT offers packet data capabilities (DECT packet radio service (DPRS)) with data rates from 24 kbit/s to 2 Mbit/s. Also part of the DECT standard are specifications for IP support and data remote access over PPP. The first data products will appear in 2000.

DECT's spectrum access method is TDD/FDMA/TDMA based on 10 frequencies each carrying 24 time-slots, providing a total of 240 channels (dynamically shared between downstream and upstream) in the 20 MHz band allocated for DECT. High data rates are obtained by allocating several time-slots (up to 23) to a link. No frequency planning is required before deploying DECT equipment because the band is reserved and because DECT is designed to share the spectrum efficiently with other DECT systems nearby. However, the co-existence of densely packed data-enabled systems has not yet been proven in practice. DECT allows for network mobility and roaming between cells of coverage which range from 50 m indoors to 750 m in an outdoor line-of-sight situation using directional antennas.

As the DECT standard also includes interworking standards with ISDN, GSM and IP networks, DECT can be used as a common access mechanism to these networks. DECT is a recognised IMT-2000 technology and could therefore play a role in 3rd generation networks and UMTS.

### 8.3.3 Wireless LAN

Wireless LANs (WLANs) use the licence-exempt frequency band between 2.4 GHz and 2.483 GHz (ISM band) which is available world-wide. 2.4 GHz WLAN products provide a wireless extension to wired customer data networks (e.g. Ethernet, token ring) and are designed to easily integrate with them. The range of WLAN applications includes indoor applications, peer-to-peer networks and outdoor bridging between buildings.

The ISM band used by WLAN is shared with other radio systems. Devices operating in that band are bound by certain rules to allow the co-existence of multiple systems but are free to be deployed wherever required. To combat interference, WLAN systems use two different spread-spectrum technologies — direct sequence (DS) and frequency hopping (FH). Since July 1997, the IEEE802.11 standard exists to allow interoperability of products from different manufacturers using the same spread-spectrum technology. In spite of the development of the IEEE802.11 standard, interoperability is still an issue.

The acceptance of WLAN is growing as the performance increases and as systems are targeted to the right customers with the right needs for cordless technology. The current price places them out of reach for residential users.

Today WLANs offer typical air-rate throughputs between 1 and 11 Mbit/s (0.5 to 6 Mbit/s user throughput after signalling is taken into account). Like all radio technologies, throughput in a cell is shared among all the users of the access point. The range of WLAN systems can be adapted to different use by employing different antennas going from omnidirectional to very directional. Typical ranges for WLAN systems are from 50 m to 200 m using omnidirectional antennas, but the range can go up to 5 km using directional antennas.

Congested areas with many users and heavy traffic load per unit may require a multi-cell structure. In a multi-cell structure, several collocated access points illuminate the same area to allow for load balancing. Mobile stations within the common coverage area automatically associate with the access point that is less loaded.

Interference can also occur from neighbouring WLAN systems on other networks. These systems will be using different encryption key and frequency band (for DS systems) or frequency hopping sequences (for FH systems) and should not interfere. In the case of FH systems, manufacturers claim that around 20 different networks can coexist in the same area before seeing performance degradation.

The next step for WLAN is to move to a different frequency band in the 5-GHz area (IEEE802.11a). This will allow for higher bandwidth products in a less congested area of the spectrum. The first examples of this next generation product should appear in 2001.

### 8.3.4  HIPERLAN

HIPERLAN is a high-performance WLAN standard, aimed at multimedia applications. It has been designed by a committee of researchers within ETSI, with strong vendor influence, and is quite different from existing products.

The HIPERLAN protocol uses a channel-access mechanism based on packet-time-to-live and priority. This means that in overload situations, real-time services are prioritised over non-real-time services. Load is also reduced by dropping packets which have been queued for a long time and can be discarded (e.g. real-time voice or video packets). For example, video will be protected at the expense of lower priority traffic such as file transfers. HIPERLAN/1 works in dedicated spectrum (5.15 to 5.35 GHz, in Europe). The signalling rate is 23.5 Mbit/s, and five fixed channels are defined.

The ETSI broadband radio access networks (BRANs) project is currently working on the HIPERLAN/2 standard. Whereas HIPERLAN/1 defines a specification for wireless access to computer LANs such as Ethernet and token ring, HIPERLAN/2 will provide local wireless access to different broadband infrastructure networks such as IP, ATM and UMTS. It will provide the quality of service (QoS) that users expect from a telecommunications network. HIPERLAN/2 will offer signalling rates up to 54 Mbit/s and will be capable of supporting multimedia applications.

No HIPERLAN products exist today, largely due to component costs at these high frequencies. The announcement of several initiatives to develop products in the 5-GHz band (HIPERLAN, IEEE 802.11a) is likely to bring costs down and make HIPERLAN products a reality. HIPERLAN's killer application might come from the move to provide a high-bandwidth wireless distribution system for digital video in the home.

### 8.3.5  Bluetooth

Bluetooth utilises a short-range radio link to exchange information between mobile telephones, mobile PCs, hand-held computers and other peripherals. The radio will operate on the globally available licence-exempt 2.4 GHz ISM band (like WLANs), allowing international travellers to use Bluetooth-enabled equipment world-wide.

Bluetooth will eliminate the need for business travellers to purchase or carry numerous, often proprietary cables by allowing multiple devices to communicate with each other through a single port. Enabled devices will not need to remain within line-of-sight, and can maintain an uninterrupted connection when in motion, or even when placed in a pocket or briefcase. Bluetooth also promises to deliver rapid ad hoc connections, and the possibility of automatic, unconscious, connections between devices.

Bluetooth uses very fast frequency hopping (1600 hop/s) to avoid interference in the ISM band. Bluetooth supports voice and data at speeds of up to 721 kbit/s and a range of up to 10 m with 100 m as an option. Bluetooth chips are designed to be incorporated into a variety of devices. They will consume very little power and manufacturing costs are expected to be between £10 and £15, dropping to £3 by 2001. The low cost should encourage manufacturers to integrate it into their products as standard, as is done with infra-red IrDA today. Bluetooth's ubiquity and the possibility of unconscious networking will allow the design of novel products and services.

First end-user products are expected to appear in the summer of 2000 in small quantities and at relatively high cost. It is likely that mass-market volumes and prices will not be achieved before the end of 2001. The real challenge for Bluetooth will be in the interoperability of devices which, if it should be achieved at the radio level, will also need to be achieved at the application level. This is more of a tall order and only the future will tell if manufacturers will choose to develop compatible products or will want to differentiate their products as much as they can.

### 8.3.6  Customer Network Integration

Cordless technologies offer interfaces which make it easy for them to be connected to data networks. DECT provide interfaces to PSTN, ISDN and LANs and is designed to carry IP traffic as native IP or over PPP. Wireless LANs and HIPERLAN both offer interfaces to LANs and offer native IP support. Bluetooth specifies serial interfaces to integrate with network terminal adapters and supports IP over PPP. Bluetooth native IP support is planned for the next version of the specification at the end of 2000.

For non-LAN-based customers, all the technologies mentioned above support dial-up networking to allow wireless modems to be built. With the launch of ADSL, cordless technologies will offer interesting possibilities to provide a wireless connection to ADSL modems so that broadband services can be distributed wirelessly in the home. For LAN-based customers, integration with WLAN or HIPERLAN is trivial. Integration with DECT or Bluetooth necessitates appropriate access points with Ethernet interfaces which are not currently available.

In all cases, the wireless nature of the link is transparent to the user.

### 8.3.7 Localised Cellular Access

Spectrum allocated to cellular operators is limited and this can cause a problem in areas with a high density of population such as city centres or inside buildings. In some areas, there can also be a coverage issue where the cellular signal is shadowed by buildings or walls. Moreover, despite the efforts to offer higher data rate radio interfaces to the cellular network (see section 8.2.1), the raw capabilities of the cellular radio will never be able to achieve the data rates which cordless technologies can achieve.

One solution for the cellular operator is to provide localised access to its cellular network using micro-base stations. Several solutions already exist based on GSM technology. DECT and soon HIPERLAN/2 both specify interworking with the cellular network so that the cordless technology can be used as a radio access mechanism to the cellular network instead of using GSM radio.

HIPERLAN/2 or DECT could be used in addition to UTRA TD/CDMA (see section 8.2.3) for wireless access to the fixed part of UMTS to provide indoor broadband data services where cellular UMTS capabilities are not sufficient. Both DECT and HIPERLAN/2 standards are being drafted so that they can operate with the UMTS network.

Technically, using cordless technology to interface with the cellular network is a cheaper alternative to offering microcells using GSM. However, the UK regulation does not currently allow the use of DECT for unlicenced public telecommunications service. The situation for HIPERLAN is currently under review in the UK by the Radio Agency (RA). This limitation can be overcome in the business/domestic environment by providing the microcell service as a managed service where the equipment is owned by the user and is connected to the operator over private circuits.

### 8.3.8 Regulatory Issues

#### 8.3.8.1 DECT and ISM Bands

Systems that operate in the DECT band require type approval to ETSI DECT requirements CTR6, CTR10 and CTR22. These requirements specify the behaviour of DECT systems so that they can share the same spectrum efficiently and with minimum interference.

Systems which operate in the 2.4 GHz ISM band require type approval to ETS 300 328. This specification requires that systems use spread-spectrum techniques and limits the transmitted power to 100 mW. Bluetooth and IEEE802.11 wireless LANs fall under these requirements.

### 8.3.8.2  Private Versus Public and Applicable Licensing Regime

Licensing regimes vary throughout Europe and the world. In the UK, the Radio Agency gives a clear definition of public and private radio systems, each with their own wireless telegraphy act licensing regime.

### 8.3.8.3  Private

A private mobile radio system is a self-provided and self-used mobile radio system. In this context, 'self' may include partners and contractors working for the licencee. Every individual has a self-provision licence by default and does not have to apply for it.

### 8.3.8.4  Public

A public mobile radio system is a mobile radio system provided commercially for use by others. Telecommunications operators must apply for an individual Telecommunications Licence, but bodies such as hotels and airports can provide commercial wireless services to customers on their premises under the Cordless Class Licence (CCL). These bodies need not apply for this licence — the service is deemed to be automatically licenced if the conditions of the class licence are met. However, the CCL only applies to DECT technology and the now largely defunct CT2 standard.

### 8.3.8.5  Licence Exemption

Use of equipment in so-called 'licence-exempt' bands (DECT or ISM) does not require a licence if and only if the type of use of the equipment is for private self-provided communications. The exemption does not extend to systems forming part of a public system which may either require individual licensing, or may be considered inappropriate for deployment in such areas of spectrum (i.e. the traffic is deemed public even if it does not connect directly to the customer but instead forms part of the system's supporting infrastructure).

Licence exemption also applies with DECT systems only under the CCL for companies providing a service to their customers on their premises. The application of the CCL to the ISM band is currently under review in the UK.

## 8.4  Fixed Wireless Access Developments

Fixed wireless access (FWA) systems are used to connect customers to the network operator's point of presence (typically a local exchange or DP) when it is not economically or physically possible to access the premises with cables. The use of point-to-point radio links has long been a useful option for connecting high-bandwidth customers where existing wired access network infrastructure is not available — however, the costs of such solutions are high. New point-to-multipoint architectures, similar in concept to cellular mobile networks, allow a sharing of hardware and bandwidth costs between users in a 'cell' and hence reduce costs to a point where a wider market deployment is possible.

The frequencies used for FWA systems are generally higher than mobile or cordless access. Two implications are a reduced maximum radio range, and the requirement for a line-of-sight or near-line-of-sight path from the base station to the customer radio unit. The net result is reduced cell radius in practical environments, ranging from as little as 1 to 3 km for wideband systems in built-up areas with unfavourable weather conditions, through 3 to 5 km for typical narrowband deployments, up to 15 km for the most ideal situations.

There are a wide range of FWA systems available, from very low cost wireless local-loop products, offering voice connections for developing countries, up to high-bandwidth LMDS systems. Here the most significant offerings aimed at extending the reach of an IP core to the edge of a customer's premises will be introduced.

### 8.4.1 Narrowband FWA

Narrowband IP-enabled FWA systems are available operating in licensed spectrum from 2 to 10 GHz and also the unlicensed 2.4 GHz ISM band. Data rates to end customers are increasing over time with current figures being around 1 to 2 Mbit/s peak.

#### 8.4.1.1 2.4-GHz Band

The ISM band products tend to be the lowest cost, but offer the lowest product and service quality. As previously stated, the band is also used by high-volume, low-cost consumer radio products — WLANs. Several manufacturers, predominantly US start-ups, have decided to exploit this by using the same low-cost radio chip-sets to make FWA wireless IP products.

These systems are designed to be used by operators for offering commercial service, but tend not to have the level of quality and network integration demanded by larger operators.

There are several products in this band which are simply 3.5 GHz systems with different frequency radio units. These are higher cost, as they are not based on the volume WLAN radios, but generally are better engineered carrier-grade products. They may also be circuit, rather than packet based, and offer IP only as an addition to their primary interfaces, such as POTS and ISDN.

### 8.4.1.2  2.5-GHz Band

Currently only available in the US, the multipoint microwave distribution system (MMDS) operates at 2.5 GHz. It is unlikely that this system could ever be deployed in many areas in Europe since the frequency band is currently allocated to NATO forces (with the exception of Ireland). This technology has evolved from US local television distribution systems, and the latest products now offer full IP support aimed at small businesses and high-end residential users. From an operator's perspective, a key attraction of these services is their range — in some areas of the US mid-west, systems are known to be operating over ranges in excess of 16 km.

### 8.4.1.3  3.5-GHz Band

Following on from the older circuit-based products, such as that used by Ionica in the UK in this band, there are now a range of IP-capable systems emerging at 3.5 GHz. This band is particularly important as it is being licensed for the use of such systems by many administrations, especially within Europe. Older systems, using circuit-switched connections between the base station and each user terminal are generally capable of offering IP connection, but in a very inefficient manner. Newer offerings are packet based, allowing much better utilisation of the available bandwidth, especially for bursty traffic. Some of these systems send IP packets directly over their air interface, giving maximum efficiency, but limiting their ability to support circuit-switched voice services. A common compromise solution is to encapsulate the IP for transport over an ATM air interface. The packet nature of ATM allows for efficient carriage of IP, but it has much better support for legacy services than the 'native' IP solutions.

Increasingly, however, systems are being offered with only IP ports (and voice over IP in the future). The primary reason for this is cost. The transport of circuit-switched services, such as POTS which is analogue and highly delay sensitive, requires considerable additional complexity in the base station, radio interface and, most importantly, the customer unit. Additionally, most regulators currently require POTS to include the provision of life-line services. As the radio system is not connected back to the local exchange by wire, the usual line powering of customer equipment is not possible. This means that such services require battery back-up at every customer installation for powering both the telephone and radio transceiver equipment in the event of a power cut. These factors greatly increase the cost of a highly price-sensitive product and mitigate against its introduction.

### 8.4.1.4  5-GHz Band

As mentioned previously, HIPERLAN/2 can deliver IP-based services at data rates up to 50 Mbit/s. While the initial market for this is expected to be in-office mobility, the basic technology could also be used for fixed wireless access, and the likelihood of this happening within the UK will become much clearer following the completion of the government's current consultation process on this band. Clearly, though, the combination of low-cost equipment (HIPERLAN's target price is the same as today's top-end 2.4 GHz wireless LANs) and abundant spectrum (there is twice the spectrum available in this band than there is for UMTS) could make HIPERLAN-based wireless access an attractive proposition.

### 8.4.1.5 Broadband FWA and LMDS

The term 'broadband fixed wireless access' is used to describe systems operating in the 10 to 45 GHz region and offering rates from 128 kbit/s up to 25 Mbit/s to each customer. Several different names are used to cover this technology including local multipoint distribution system (LMDS), local multipoint communications system (LMCS), broadband wireless local loop (B-WLL) or multimedia wireless system (MWS).

LMDS was originally intended to provide local television services in America. Large amounts of LMDS spectrum around 28 GHz were auctioned in the US and, due to the flexibility in licensing regulations, this spectrum could also be used to deliver general telecommunications services. Focus has shifted to the use of LMDS for data services such as IP access and delivery of leased lines. The use of the term has expanded beyond the US to refer to any high-bandwidth point-to-multipoint fixed radio system.

The majority of available systems are truly multiservice, normally using ATM for the transport of data over the air interface and back into the core network. Rates offered to individual customer's units range from 128 kbit/s to 25 Mbit/s, although, due to their cost, they are often shared between multiple tenants in the same building.

The frequencies at which they operate make these systems strictly line of sight and sensitive to heavy rain fall. To minimise these problems, the base station needs to be located as high as possible and customer antennas mounted on or above roofs. Additionally cells' radii are limited to between 1 and 3 km, depending on climate and topography.

The frequency of operation, speed, range of interfaces and quality of service offered by these systems mean they are considerably more expensive than narrowband offerings. Typically targeted at small-to-medium enterprises, but, with suitable in-building distribution, it can be targeted at multi-tenanted offices or residential buildings. The economic viability in a given region is a direct function of the density of suitable customers and limits its usefulness to urban centres.

Systems using IP directly over the air interface are being developed, but are less advanced than the simpler narrowband systems.

### 8.4.2 Optical Links

Systems using infra-red technology are available that offer data rates in excess of 100 Mbit/s. These systems provide a point-to-point link between buildings and could be used to provide telecommunications services to a customer with high-bandwidth requirements. In a similar way to wideband fixed radio systems, but to a much larger extent, rain and fog affect them. The systems require no spectrum licence, but do have to be 'eye-safe', leading to low-transmit powers. These factors place strict limits on either the achievable range or link availability. To meet the availability needs of a business customer, a typical system would only operate over less than 200 m in a city such as London.

## 8.5 Solutions for Wireless Access to a Carrier-Scale IP Network

There is a wide range of possible technical solutions to support the carrier-scale IP network. As shown in the previous sections, these solutions address two needs:

- for mobility as addressed by cellular and cordless technologies;
- for an alternative to wired network infrastructure as addressed by fixed wireless access technologies.

Fixed wireless access offers a range of alternatives to the various wireline systems for accessing customers' premises. Systems from low-speed dial IP replacement for residential users, up to multi-megabit systems offering a range of services to medium enterprises are available.

For new operators, such as BT's alliance partners overseas, establishing access to their customers is a key issue. As well as the cost, installing a fixed network takes a long time and digging the streets may not be allowed at all. Here fixed wireless access offers advantages in terms of ease and therefore speed of installation. New customers can be connected within hours of a service request, provided a suitable base station is in place. Another option for new operators is leased access from the incumbent, a much more cost-effective option than building owned wired access, at least in the short term, but less flexible — finding the right partner can also help in that there is often capacity and infrastructure available. Again wireless access offers advantages, reducing the ongoing costs and allowing end-to-end ownership of the customer. While the above makes it sound like an ideal solution to a new operator's needs, the economics are such that fixed wireless is an alternative to, rather than a replacement for, the other options. Payback times versus leasing or building wired networks can be long, even in the more densely populated areas and in some places it will never be cost effective.

In the UK, BT has an extensive and growing fixed access network and therefore the economics of fixed wireless are very different. Here, to make deployment worthwhile, the new technology needs to be cheaper than the existing options, such as fibre, copper or point-to-point radio. It still has the potential for faster service provision, but this is less of an issue for an incumbent operator, as they will already have the required access medium installed to many customers. The government has announced their intention to license suitable spectrum for both narrow and wideband systems and BT is considering its position.

In addition the success of the technology to address the need for mobility and the interworking between cordless and cellular systems will depend upon the terminal that is manufactured.

A simple configuration for UK users to access the carrier-scale IP network could be:

- users have both a cellular telephone and a wireless-enabled PDA or laptop;

- dual mode cellular handsets (GSM and UMTS) are used for data and voice in the wide area;

- the PDA is used in corporate buildings and the home via Bluetooth or another higher specification cordless technology;

- the PDA talks to the cellular handset via Bluetooth;

- parts of the network are provided by FWA solutions.

## 8.6 Conclusions

Overall this chapter has illustrated the enormous growth in opportunities to deliver services through wireless technologies.

Voice will continue to grow and will clearly move from a service that is related to a device wired to the wall and therefore specific to a location, to a service that is personal with the vast majority of people having a mobile telephone.

Data is then the growth area that through the introduction of IP and the massive rise in computer literacy will allow business opportunities that will create a vast, fast-moving market-place. How will this market grow and where will the key revenue be? How will operators recoup the money they have spent on 3rd generation spectrum? It's going to be very interesting.

Data and a range of new applications will be key to new revenue. Revenue sharing will be the new model as the relationship between wireless operators and wireless customers changes to one where content providers and value-add application developers also play a role and take a share of the revenue. It is happening in Japan and will happen across the globe in the next three years. A substantial step forward is to examine the developments of GPRS for both services and charging principles. In the UK we need to leverage on the GPRS activities and introduce systems and services that can be migrated to 3G as that comes on line.

## Reference

1   NTT DoCoMo Web site — http://www.nttdocomo.com/

# 9

# DIAL ACCESS PLATFORM

## J Chuter

## 9.1 IP Dial-up Service Components — Introduction

A significant amount of the Internet's success can be contributed to the industrialisation, the creation of a carrier-scale solution, of dial access and its associated modems. It is the mass production of modems, the significant reduction of their cost, and the development of network systems which can handle hundreds of thousands of customers, that has brought the cost of Internet access within the reach of the majority of the population. This chapter introduces the major elements of a carrier-scale dial access platform and describes how they should be organised to deliver service.

The basic components for an IP dial-up service comprise:

- modems;
- terminal server;
- authentication system.

The modems act as the interface into the telephony network and provide asynchronous data streams into a terminal server function. The terminal server encapsulates the asynchronous data into IP packets for delivery over the IP network. Since this is a remote access system, some form of user authentication is normally a requirement of the overall service. In simple networks, authentication may be carried out by the host itself as shown in Fig 9.1.

The advent of public dial-up services to the Internet, by Internet service providers (ISPs), has stimulated developments in protocols and hardware integration. The banks of modems and terminal servers have been replaced by integrated network access servers (NASs). These are fitted with digital modems which exploit the fact that the telephony network operator has deployed a digital network and that analogue/digital codecs are available in the exchanges. The interface to the NAS is typically a 30-channel ISDN primary rate interface (PRI). The widespread use of the point-to-point protocol (PPP) [1] allows many more sessions to be handled as the relatively processor-intensive task of IP encapsulation has been moved to each

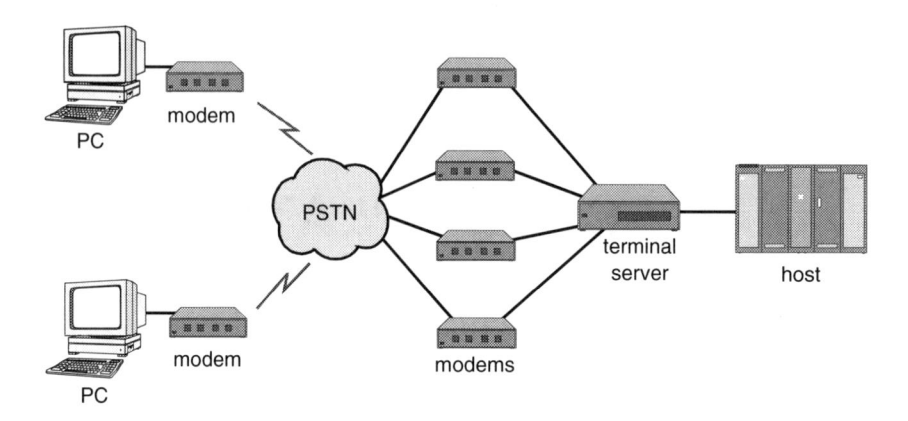

**Fig 9.1** Basic dial-up network.

individual dial-up client. Authentication has also been addressed in the standards arena with the development of the remote authentication dial-in user service (RADIUS) [2].

RADIUS is a protocol that allows a NAS to pass details of a user's identity to a server. The server performs the username/password authentication and can instruct the NAS to permit the connection or terminate the call. RADIUS is more than just an authentication protocol as it also provides authorisation and accounting functions, the latter being the subject of a separate standard [3]. A typical integrated dial-up network is shown in Fig 9.2.

## 9.2 Evolution of Network Provider Role

As the ISP grows, more telephony network connections are required. One strategy is to deploy NASs in a number of points of presence (PoPs) across the country. This allows the use of standard local rate numbers to attract users, but there are increased costs associated with running several dispersed sites as well as the need for an increased number of WAN back-haul links to connect users to centralised services, including authentication. An alternative solution is to grow a single site, making use of non-geographic numbering schemes (e.g. 0845) to provide equivalent local rate access.

ISDN PRIs are generally only available from certain exchanges within the telephony network and a large PoP can exhaust the capacity. Building dedicated exchanges or long-lining from an adjacent exchange are possible options but increase the per-port cost. This approach is unlikely to scale for very large networks as there is not the telephony capacity in a single location.

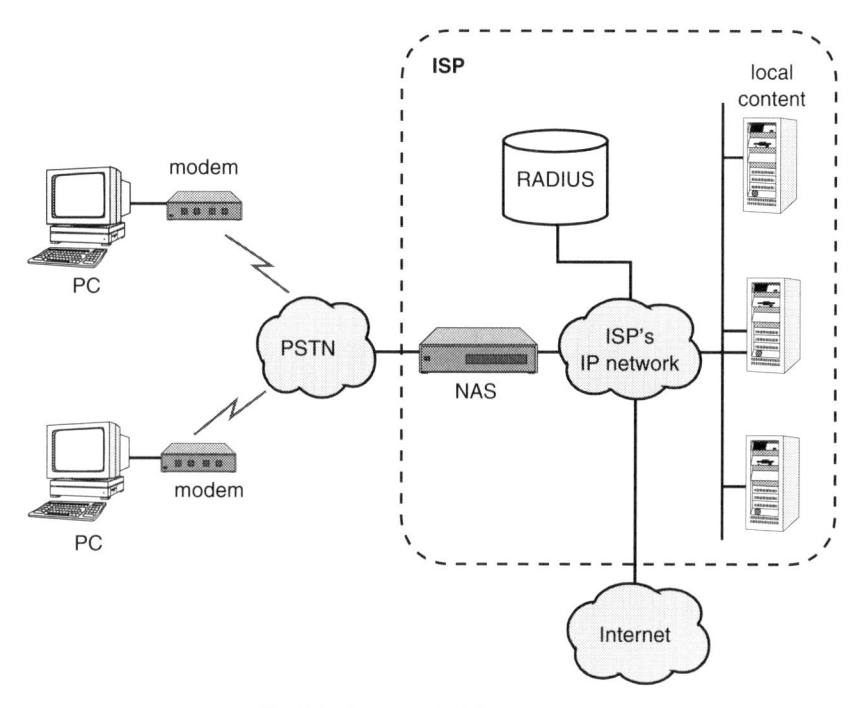

**Fig 9.2** Integrated dial-up network.

In response, some telephony network providers are providing an IP access service as a means of protecting their telephony network while improving the overall scalability and reducing costs. The use of non-geographic, typically local call rate, telephony access numbers makes the actual location of the PoPs largely irrelevant and allows the network provider to choose the location of the PoPs to match available telephony network resources. Aggregate IP host links then connect the ISPs to the telco's IP backbone network. As the costs come down, there is a steady shift of ISPs from owning/managing their own access networks to outsourcing to a telco-based solution. This outsourcing of dial-up capability is also attractive to the increasing number of corporates who require remote access to internal systems but do not want to be in the business of operating the NASs and their associated telephony links.

BT's public dial-up network encountered these same issues. Initially two PoPs were built, as much for resilience as for capacity. As the sites continued to grow, it became increasingly difficult to obtain sufficient PRI connections to sustain the expansion. Furthermore, as the ISP was being operated within BT's Supplementary Services Business (SSB) licence, the cost model favoured procurement of more 'centralised' PRI even though this was causing significant engineering difficulties within the main PSTN operated within BT's Systems Business (SB) licence.

On 1 April 1998, IP transport services moved into SB as a means of meeting the growth demands of a number of UK ISPs (not just BT's own), while maintaining the integrity of the underlying telephony platform. With the introduction of new tariffs to reflect the new architectures available, the aim was to reduce the overall per-port cost.

Further cost reductions become possible by sharing the dial platform infrastructure between a number of customers. In particular, sharing the NAS would be advantageous, as it is the digital modems that account for a large proportion of the overall costs. The degree to which this can be achieved depends in part on the type of service the customers are expecting. A corporate is likely to be expecting total privacy with no connection to the Internet. Since there is no connectivity to the Internet there is no need to use registered IP addresses and the attendant difficulties of obtaining this increasingly scarce resource from RIPE are avoided. Conversely an ISP will need to use registered addresses for its clients which can be routed over the Internet. A number of ISPs could, in theory, be served by a common platform and share the same client IP address space. In practice, however, ISPs not only facilitate access to the Internet but they also provide additional services, such as e-mail and news, and specific content. The simplest method of restricting access to these value-add services is to filter on source IP address. As a result many ISPs require their own dedicated IP address ranges and the scope for sharing the infrastructure is reduced further.

The solution is to provide virtual private networks (VPNs) over a shared platform infrastructure. This requires a step change in technology and increased complexity, but increases the flexibility of the platform to handle a much greater range of customer requirements.

## 9.3 VPN Technology

The current VPN technology for dial platforms is centred on the use of IP tunnels. There are a number of vendor proprietary tunnel protocols (e.g. PPTP, ATMP, L2F) but recently the Internet Engineering Task Force (IETF) layer-2 tunnelling protocol (L2TP) has become standardised [4] and implementations are becoming available from the vendor community. Essentially L2TP encapsulates an IP packet with another header to indicate the destination point. Transit routers do not need to inspect the original packet, which can thus include IP addresses that would be unrouteable on the backbone (i.e. are within a VPN). All tunnel protocols necessarily have two end-points. In the L2TP specification the tunnel originator (NAS function) is termed the L2TP access concentrator (LAC), while the device that terminates the tunnel is termed the L2TP network server (LNS) (see Fig 9.3).

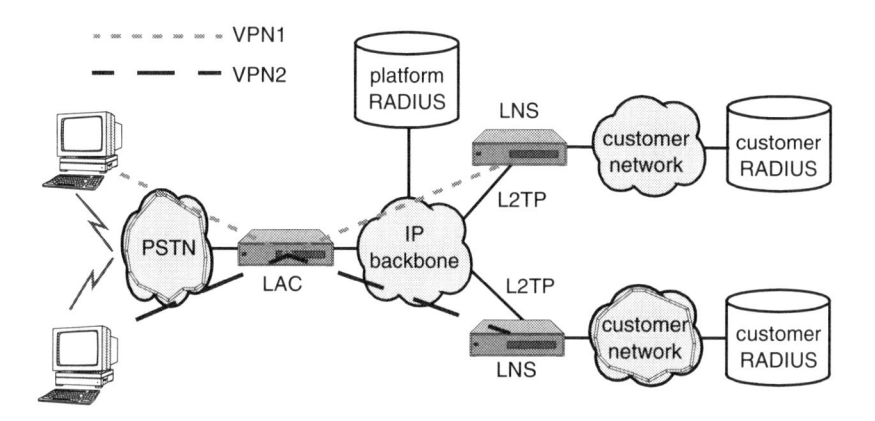

**Fig 9.3** Dial VPNs using L2TP.

L2TP performs end-point authentication prior to establishing a tunnel. On a shared platform, whereby a LAC could be building tunnels to a number of LNSs, there needs to be a mechanism of instructing the LAC to associate a particular dial-up session with a tunnel and to specify what username and password combination to use in creating the tunnel. For a scalable solution, these parameters cannot be locally configured on the LAC and the solution is to query a suitable server on a call-by-call basis. As RADIUS is already a defined AAA system within the dial platform, extensions have been proposed to the RADIUS authentication standard to allow a RADIUS server to download the tunnel attributes on a call-by-call basis. In a completely tunnelled service environment the platform RADIUS is not authenticating users as such, but is used to identify services. Typically this is done on the basis of the dialled number.

The platform RADIUS server provides the key service control mechanism. The existing standards, and extensions, provide the basis for this functionality. BT has developed a high functionality RADIUS server implementation which is specifically optimised for use in the platform server role, where service identification, and operation on a shared platform with multiple customer systems, are more important than the ability to handle end-user accounts.

## 9.4 Platform Scaling

The dial platform comprises a number of component elements, each with their own scaling issues and solutions:

148  *Platform Scaling*

- telephony delivery;
- PoPs;
- RADIUS;
- backbone network;
- tunnel termination.

### 9.4.1 Telephony Delivery

In a conventional network, ISDN PRIs are available from specific exchanges — in the BT network these are digital main switching units (DMSUs). There are a limited number of these exchanges and the associated interface cards. As the dial platform becomes larger it becomes increasingly difficult, and thus costly, to find and provide enough interface capacity.

Vendor LAC implementations are essentially customer premises equipment (CPE). The telephony signalling standard for CPE is Q.931. Within the main telephony network the signalling standard is SS7 and one of the functions of the exchange ISDN PRI line card is to perform SS7 to Q.931 translation.

The use of an alternative translation device would eliminate the need for ISDN PRI line cards and opens up the possibility of using any exchange to source traffic into a PoP. The PoPs can then be deployed closer to the origination point to minimise the amount of data traffic being carried over the telephony trunk network. In the limit, each local exchange could be directly connected to a collocated PoP (data offload).

A number of vendors offer an SS7 signalling gateway to interwork with their LAC. The signalling gateway converts SS7 messages into Q.931 and forwards them to the LAC over an IP network. The use of IP encapsulated Q.931, known as Q.931+, makes good use of existing infrastructure but introduces additional constraints on that infrastructure with respect to latency. The SS7 gateway is typically a powerful computer to handle the real-time, processor-intensive translation task on a large scale.

The signalling gateways will tend to be relatively few in number as current generation systems can handle 10 000 or more simultaneous sessions. As more PoPs get deployed, the signalling network becomes increasingly complex to deliver the SS7 circuits from across the country to the signalling gateway sites. Implementations of SS7 over IP are becoming available and the way forward may be to locate an SS7 encapsulation device in the PoP and make use of the existing IP infrastructure to transport the SS7 messages.

These encapsulation devices can handle the lower layers of the SS7 protocol stack and thus relieve the signalling gateway of the most processor-intensive operations, allowing a greater number of sessions to be managed by a single system. In effect, the PoPs become remote processor units for the signalling gateways, as shown in Fig 9.4.

BT has already deployed an extensive Q.931+ based network as part of its overall dial platform and is well advanced in the design of a practical, scalable SS7-over-IP solution as described earlier.

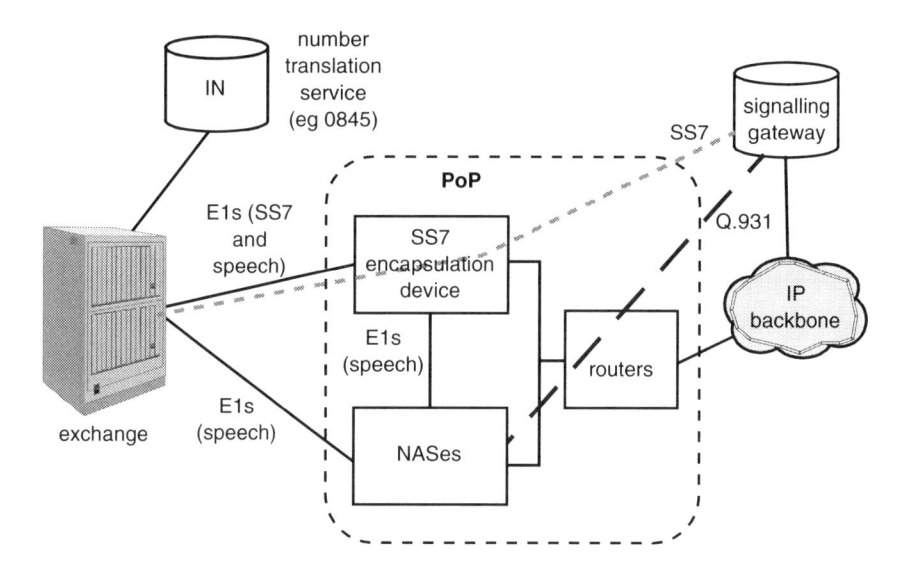

**Fig 9.4** Integrated, IP based signalling.

### 9.4.2  PoPs

Unlike other components within the platform, the internal PoP design does not have to scale up to the same extent. Indeed, as the platform moves towards a highly distributed PoP environment, there will be a tendency to deploy small PoPs and the main scaling issue is how these remote sites connect to the IP backbone network. The dial PoPs are too numerous to be directly connected to the IP backbone network and instead an access layer will need to be introduced.

150  *Platform Scaling*

For a telco solution the backhaul network is most likely to be ATM based. The PVCs from the dial PoPs are aggregated on (local) access routers which will forward traffic into the main backbone network. A key design factor is the average bandwidth per dial port. This will be much less than the modem line rate (i.e. contended access), but will still yield links of several tens of megabits for PoPs of a few thousand ports.

### 9.4.3 RADIUS

The RADIUS platform provides the mechanism for identifying which service a user is trying to reach. As the number of ports increases, so too does the volume of RADIUS transactions. From a maintenance and environmental perspective, grouping servers together makes sense and so scaling the RADIUS platform is mainly concerned with designing server farms (see Fig 9.5. Typically the hardware will be Unix based, and a scalable database, such as Oracle, will be used.

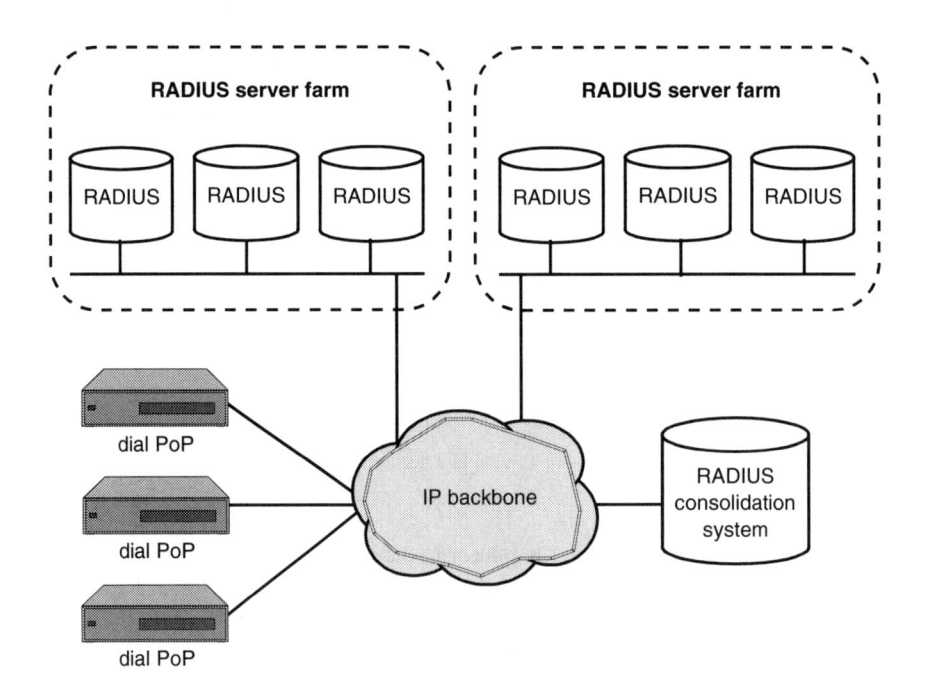

**Fig 9.5**  RADIUS server farms.

Physical location of the server farms becomes an important factor as computer-hall-style accommodation will be required to host the servers with good IP connectivity to the PoP sites. RADIUS traffic is not very large in terms of raw bandwidth, but low latency is needed to reduce the authentication time. These requirements are the same as for the signalling gateways, and indeed other network services such as DNS.

The platform RADIUS servers are responsible for identifying services not users. The rate at which services are churned (added/deleted) is likely to be quite low and there is little need for real-time database updates. A master database can be modified and replicated to the distributed databases on a periodic basis (e.g. once a day). Accounting records need to be collected from each of the machines and consolidated. Start records for a session can be delivered to one machine while the stop record is delivered to another. The RADIUS protocol is connectionless as it uses UDP; therefore this split accounting record situation can occur under normal conditions if a load-balancing device is used within the server farm, or under load conditions if multiple servers have been configured in the LAC. Typically the consolidation function would be carried out on a separate box in non-real time and would remove duplicate entries (retry attempts) as well as matching corresponding start and stop records.

Real-time updates between databases are needed to perform a concurrency checking function. Customers (i.e. ISPs and corporates) typically purchase a number of dial ports from the network provider. This equates to simultaneous sessions for each customer.

The concept of sharing infrastructure necessarily means that there will be more physical ports than any individual customer has ordered. Potentially a customer could establish more simultaneous sessions than ordered.

In some cases, commercial terms, such as an excess charge, may be appropriate, but these 'controls' can only be applied after the event. In many situations the requirement will be to prevent a user logging on to the network. In some cases this can be achieved by simply restricting the number of IP addresses available to be given out to clients on the LNS. Clearly this cannot be done if the address assignment is being done by the customer (RADIUS) or the user (static addresses). In these cases the platform RADIUS would have to perform a concurrency check to establish whether the limit had been exceeded for the customer. This check has to be done in real time and relies on updates between databases. These updates need not be the full accounting records, as only the number of current sessions for each customer is required to perform the concurrency check. The ultimate situation is to perform the check prior to answering the telephony call so that the user is not charged for the failed call.

Some vendor LAC implementations offer this 'RADIUS before answer' functionality, although questions still remain as to how best to indicate to the user, through the use of available MF telephony tones, that the service is busy, rather than the network is engaged or faulty.

### 9.4.4 Backbone Network

The backbone network scaling issues are discussed in detail by in Chapter 3. Typically the dial access platform will appear to the backbone network like a large number of fixed connections. Until DSL access really becomes established in the public networks, the dial platform will continue to be the driving force behind the expanding Internet.

### 9.4.5 Tunnel Termination

Terminating the L2TP tunnels in an efficient, scalable manner is crucial to the success, both technically and commercially, of the overall solution. The most cost-effective design is to share equipment. As discussed earlier, most customers want to use their own IP address space and utilise other features (e.g. filters, time outs) which are customer specific. This reduces the potential for equipment sharing, although virtual routers provide a possibility. The whole concept of tunnelling is to forward the PPP frames from the LAC to the LNS, which will now have to perform the PPP termination and extract the user's IP packets. This is a processor-intensive task and relatively few sessions can be handled. With demand for dial services rising, the design will have to accommodate multiple LNS working.

All user traffic will get delivered to the LNS, as this is a fundamental tunnel operation. In a corporate VPN situation, this traffic is all destined for the corporate's own network. For an ISP, the situation may be that most of the traffic is destined for the Internet and 'tromboning' the traffic through the ISP's network leads to additional expense in terms of host-link capacity and router capability for the ISP. In this case, the LNS would be better deployed within the dial network provider's accommodation to make use of available network bandwidth to the Internet with a correspondingly smaller link to the customer site for RADIUS and local ISP content.

Even for standard VPNs there is a considerable management and maintenance gain in deploying the LNS equipment for large services within the dial network provider's accommodation (see Fig 9.6). Expansion is much less obtrusive, and there is flexibility in deployment of technologies.

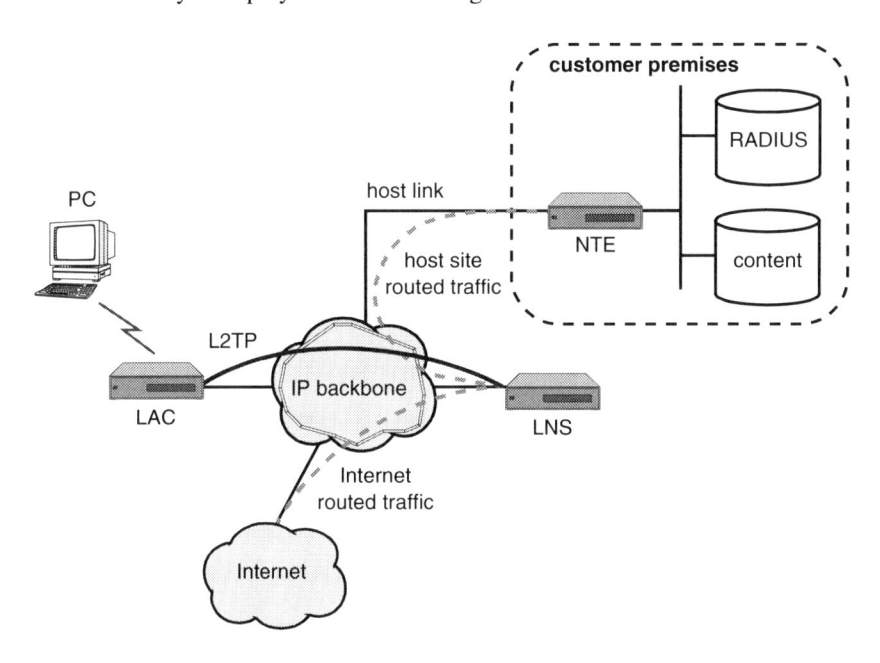

**Fig 9.6** Network provider located LNS.

In order to realise very large VPN services, a method of grouping LNSs is required. Functionally, the design for an LNS 'cluster' comprises front-end routers and a bank of LNSs. The front-end routers provide the connectivity to the backbone network and also to the customer site.

The design of these LNS 'clusters' needs to support the range of customer service options, typically RADIUS driven, and also allow expansion to multiple 'cluster' working to realise very large numbers of simultaneous sessions. Some service aspects are especially difficult in this sort of environment. These include client-assigned (static) IP addresses and handling ISDN 2B multilink PPP sessions.

154  *Standards*

The principal problem is that individual sessions will tend to be delivered to different LNS elements. For client-assigned addresses, the problem is notifying the customer's network of the changed return path in a timely fashion. For multilink PPP, there are a number of problems associated with identifying and recombining the constituent PPP sessions while maintaining a performance better than an individual ISDN B-channel. The situation can be eased by using different telephony numbers for these types of services, which allows the platform RADIUS server to issue different tunnel-build attributes to the LAC and so restrict tunnel delivery to dedicated LNSs.

The current BT product definitions for IP dial services specify a LAN interface to the customer. The availability of L2TP as a standards-based protocol opens up the possibility of delivering L2TP tunnels directly to the customer. This would move the LNS complexity into the customer domain. It remains to be seen whether customers will take up the challenge of this sort of network interface, as, although the protocol is standard, it is still immature and the vendor implementations do not necessarily interwork. It seems more likely that direct L2TP interfacing will be used between network providers as an alternative to existing IP and PSTN interconnects.

## 9.5  Standards

The IETF provides the main forum for advancing standardisation in IP-based protocols. The IETF organises a number of working groups (charters) to develop standards in particular specialist areas. From a dial platform perspective, the most significant charters are PPP and RADIUS. The PPP charter includes native PPP, multilink extensions and L2TP. The RADIUS charter includes RADIUS authentication, RADIUS accounting and tunnel support. As would reasonably be expected, the most recent publications are available over the World Wide Web [5—8].

## 9.6  Conclusions

This chapter has described the component systems that make up BT's UK IP dial platform. Scalability is a key feature of the overall design and introduces a range of additional problems. Scaling the telephony network has the greatest impact on the platform design, and favours a distributed PoP approach and the sharing of network resources between many customers. This in turn dictates the use of VPN tunnelling techniques and a novel signalling network based on emerging IP-based standards. BT has implemented a large-scale IP dial network using these advanced technologies, while still maintaining flexibility in service provision.

## References

1 Simpson, W.: '*The Point-to-Point Protocol*', RFC 1661 (July 1994) — http://www.ietf.org/rfc/

2 Rigney, C. et al.: '*Remote Authentication Dial in User Service (RADIUS)*', RFC 2138 (April 1997) — http://www.ietf.org/rfc/

3 Rigney, C.: '*RADIUS Accounting*', RFC 2139 (April 1997) — http://www.ietf.org/rfc/

4 Townsley, W. et al.: '*Layer 2 Tunneling Protocol (L2TP)*', RFC 2661 (August 1999) — http://www.ietf.org/rfc/

5 IETF home page — http://www.ietf.org

6 L2TP charter — http://www.ietf.org/html.charters/l2tpext-charter.html

7 PPP charter — http://www.ietf.org/html.charters/pppext-charter.html

8 RADIUS charter — http://www.ietf.org/html.charters.radius-charter.html

# 10

# AN OVERVIEW OF SATELLITE ACCESS NETWORKS

### M Fitch and A Fidler

## 10.1 Introduction

Satellites are raising their profile in the access network, migrating from their traditional role as trunk connections between core networks. This migration is being driven in part by fibre installation around the globe and in part by increasing demand for mobile and IP-based services to which satellites are well suited. This chapter provides an overview of those satellite technologies that can be used to deliver IP-based broadband services to customers.

Satellite communications have some unique properties when compared to the other access technologies described in this book. Current communications satellites comprise two basic types. The first type, which is by far the most common, operates only at layer 1 (physical) of the OSI model; they receive radio signals from earth in one frequency range, translate the radio signals to another frequency range, then they transmit the signals back to earth. They are therefore 'transparent' to layers 2 and above. These satellites are used for data rates from 16 kbit/s to 155 Mbit/s and all are in geostationary[1] orbit (GEO). The second type, of which Iridium was an example, operates at layers 1 and 2. These satellites translate the radio frequency as in the first type and in addition have ATM-like cell switches on-board that operate at layer 2 (link layer).

In terms of future communications satellites, proposals worth over $30B are filed with the FCC that describe around 20 'next generation' satellite systems operating in various orbits with varying quantities and sizes of satellites. Both transparent and layer 2 switching satellites are proposed, and their launch dates are spread out between 2003 and 2007. This leads to a high number of combinations and this

---

[1] Geostationary orbit satellites have an orbit period of exactly one day and are at an altitude of about 36 000 km above the equator. When viewed from the surface of the earth, the satellites appear stationary. Three are required for global coverage. Low earth orbit (LEO) satellites have a typical orbit period of 100 minutes and are at an altitude of several hundred km. Several tens of satellites are required for continuous global coverage.

158  *Introduction*

chapter does not have the space to give an evaluation of these systems. They are aimed at different areas of the market but tend to have a common theme; they are all for access. An example of a next generation GEO layer-2 switching satellite is shown in Fig 10.1, illustrating how several services can be up-linked from different geographical areas and multiplexed to a user.

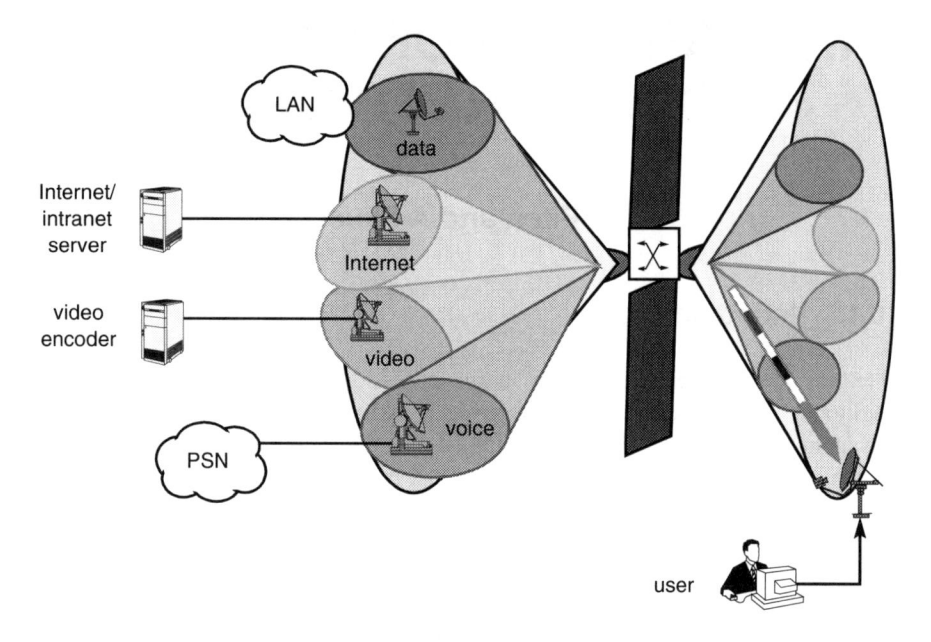

**Fig 10.1** A layer-2 switching satellite.

There are also proposals for high-altitude platforms (HAPs), which share certain characteristics with satellites; these are communications payloads fitted to balloons or aircraft flying in circles at a height exceeding 50 000 feet. They are designed as an access technology with user rates at up to a few Mbit/s. At least one prototype aircraft has been flown.

Historically, two-way communications satellites have been used to provide trunk connections, particularly to countries that are difficult to reach. There are still several tens of countries where satellites provide the only reliable means of connection. The services provided are unswitched circuits or switched (dialled) circuits that offer private leased circuits or public-switched voice circuits respectively. Of all the satellite services, these traditional ones still generate the greatest revenue and profit to the service providers and satellite operators. However, a sizable chunk of this revenue disappears whenever an optical cable is installed that removes a country from the 'difficult' list. The challenge here is that the migration rate to the access domain is not yet high enough to counter losses in trunk revenue.

However, the rate is predicted to climb steeply as the next generation of satellite systems is introduced.

Apart from trunk communications, satellites have always been used for television transmission between countries and, over the past 15 years, directly to user premises where technology and higher frequencies (Ku band) have been used to reduce the size and cost of receiving equipment. TV is now transmitted digitally using the digital video broadcast (DVB) ETSI standard. DVB, which is a layer-2 protocol, is a multiplex containing several programme channels that can be broadcast from a satellite at a gross rate of about 35 Mbit/s, dependent on the modulation scheme used. Within the DVB multiplex, it is now common practice to replace some or all of the MPEG-coded video streams with IP datagrams using multiprotocol encapsulation (MPE). This enables a mix of TV and Internet services to be delivered directly to users with a 60—90 cm antenna and low-cost receiving equipment. It is this development in particular that has enabled satellite to become cost effective in delivering Internet services.

## 10.2 Attributes and Services

From the service provider perspective, the main attributes that satellites provide are fivefold.

- Wide area coverage

    Current satellites have a variety of antenna types fitted to them that generate different footprint sizes. The sizes range from coverage of the whole earth as viewed from space (about 1/3 of the surface) down to a spot beam that covers much of Europe or North America. All these coverage options are usually available on the same satellite, selection between coverage is made on transparent satellites by the signal frequencies (i.e. at layer 1). It is spot beam coverage that is most relevant for access since they operate to terminal equipment of least size and cost. Future systems will have very narrow spot beams of a few hundred miles across that have a width of a fraction of a degree. If the satellite has layer-2 switching, then the choice of downlink beam is determined by the switch port. An illustration of such coverage is shown in Fig 10.2.

- High bandwidth

    The transponder bandwidths on communications satellites is commonly 72 MHz (on TV broadcast satellites it is 36 MHz). This has been proved adequate through trials to support STM-1 (155 Mbit/s). The problem is that large earth stations are required. However, future satellites with narrow spot beams can deliver rates of up to 100 Mbit/s to a 90 cm antenna and the backplane speed within the satellite switch will be typically in the Gbit/s range. The uplink rate from a 90 cm user terminal is typically 384 kbit/s.

160 *Attributes and Services*

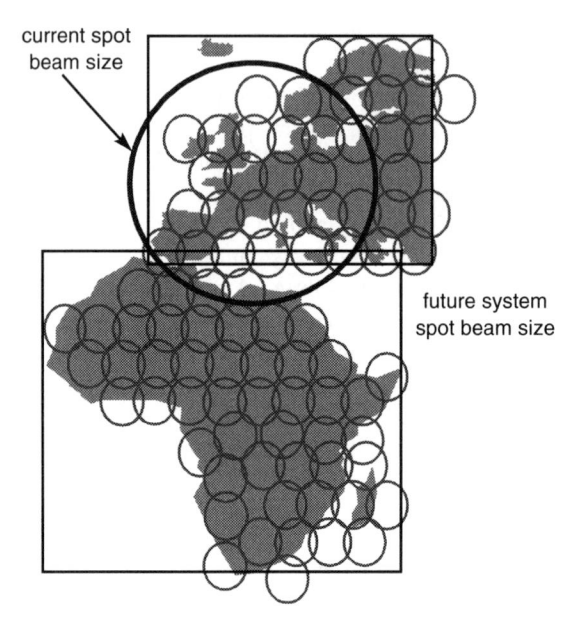

**Fig 10.2** Comparison of current and future system spot beam sizes.

- The cost is independent of distance

  The wide area coverage from a satellite means that it costs the same to receive the signal from anywhere within the coverage area.

- Fast access

  Once the hub and network connection is in place, which is necessary for the first user, more users can be added in the time it takes to install the equipment, pending regulatory requirements.

- Increased reliability and security

  Satellite links only require the end stations to be maintained and they are less prone to disabling through accidental or malicious damage.

These attributes combine to make satellites a flexible and fast way to connect users. They can be used as a strategic tool to bring services to market quickly, to bring in early revenue, ahead of the terrestrial network development. They can be used to reach outlying company and residential premises quickly in difficult and rural areas.

In terms of services, the broadcast nature of satellites combined with the 240–280 ms hop delay encountered with GEO systems makes their use more suited to some services than to others. Satellites are less suited to:

- games and other applications that require fast interaction;
- large file downloads using Microsoft TCP stacks — the hop delay limits the throughput to a maximum of approximately 1 Mbit/s on high bandwidth links due to the TCP windowing mechanism; this problem can be mitigated by using UDP and building in reliability at the application layer, by using commercially available 'satellite-friendly' TCP stacks or by spoofing — the problem does not arise with Internet 'surfing' since the browser sets up several parallel TCP sessions;
- voice with inadequate echo cancellation.

The applications listed above are affected by the GEO hop delay; however, this delay is much reduced with LEO satellites. In fact an LEO satellite hop exhibits less delay than a fibre for distances greater than about 1/3 of the way around the earth, because the speed of light in a fibre is only about 0.6 c.

Satellites are ideally suited to:

- one-way services to a large audience, such as video, audio and data streaming;
- asymmetric services such as Internet access and ISP backbone connection — the return link can be over a terrestrial link (such as ISDN) or over the satellite;
- multicast IP over a layer-2 protocol such as DVB or point-to-multipoint ATM — a layer-2 switching satellite can be used to replicate ATM cells into two or more spot beams as well as multiplexing traffic from several beams into one.

The range of services to which satellites are suited will increase as the future systems are deployed. One example will have over 200 LEO satellites with inter-satellite links between them and will be suitable for all IP access services.

## 10.3  IP Access Delivery via Satellite

There are several link-layer (layer-2) options for delivery of IP via satellite. The simplest is the use of HDLC or PPP over a serial link between an IP router and a satellite modem, as shown in Fig 10.3.

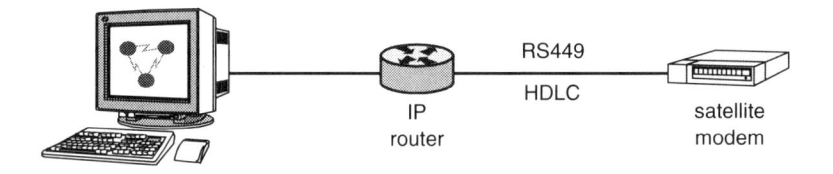

**Fig 10.3**   IP via serial encapsulation.

162  *IP Access Delivery via Satellite*

This delivery mechanism is an effective solution for commercial point-to-point services requiring LAN connectivity at the customer premises equipment (CPE) end. However, this situation becomes costly and unpractical for mass-market residential delivery of IP services. The main reasons are high CPE costs and the lack of broadcasting and multiplexing capabilities. This solution is suited to ISP backbone connections to the Internet, whereby the link can be symmetrical, asymmetrical or one-way. Furthermore the router can use policy-based rules to shape the traffic, for example by altering the available bandwidth as a function of time of day.

Another possibility is to use ATM at the link layer, employing 'classical IP over ATM' with ATM adaptation layer (AAL) 5.

Trials of ATM over satellite have been carried out by BT at Adastral Park to measure the effect of a satellite link on constant bit rate (CBR) and variable bit rate (VBR) parameters, such as cell delay variation (CDV). It was found that the CDV is less than 0.1 ms on a satellite circuit which means that CBR is feasible if the GEO hop delay (240—280 ms) can be tolerated as part of the overall 400 ms cell transfer delay budget. One such trial used one-way ATM over satellite, where the forward link is transmitted over satellite providing a bandwidth to the user of up to 8 Mbit/s and the return link uses ISDN at 64 kbit/s as shown in Fig 10.4.

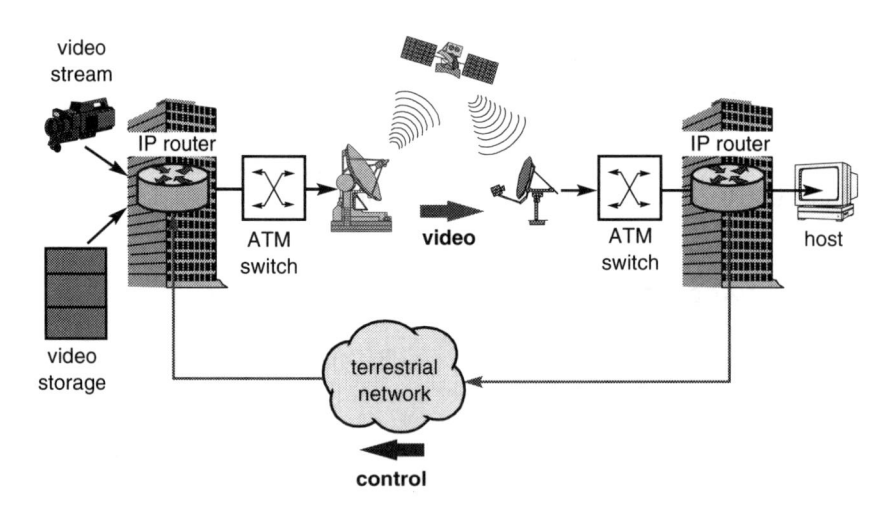

**Fig 10.4** Trial of one-way ATM over satellite.

With this trial, the virtual circuits of the ATM layer were used to separate (and hence to separately police) user traffic from network management traffic, enabling integrated network management, including monitoring of end-to-end quality of service (QoS).

An Overview of Satellite Access Networks 163

A similar arrangement could be to use frame relay as the link layer, which would use different data link connection identifiers (DLCIs) to separate network management traffic from user traffic in much the same way as ATM. There are situations where it is better to use frame relay rather than ATM at the link layer; the most compelling is where the combined bit rate from all the channels on a single link is below about 2 Mbit/s. This is the bit rate below which the multiplexing gain with ATM becomes outweighed by the overheads associated with the cell header and maintenance cells.

An example of IP services delivered over frame relay via satellite is the very small aperture terminals (VSAT) system designed to provide symmetrical access to a frame relay network for Eastern Europe. The rates offered are up to 128 kbit/s to 1.8 m C-band (4–6 GHz) antennas in this particular system; however, there are VSAT products on the market that provide up to 2 Mbit/s symmetrical links to 1.8 m Ku-band (11–14 GHz) antennas.

Another possibility is to use DVB as the layer-2 protocol and, as stated in section 10.1, it is this option that provides the greatest potential for delivery of IP services. A system that delivers IP services over DVB by satellite in the forward direction with a terrestrial return link is shown in Fig 10.5.

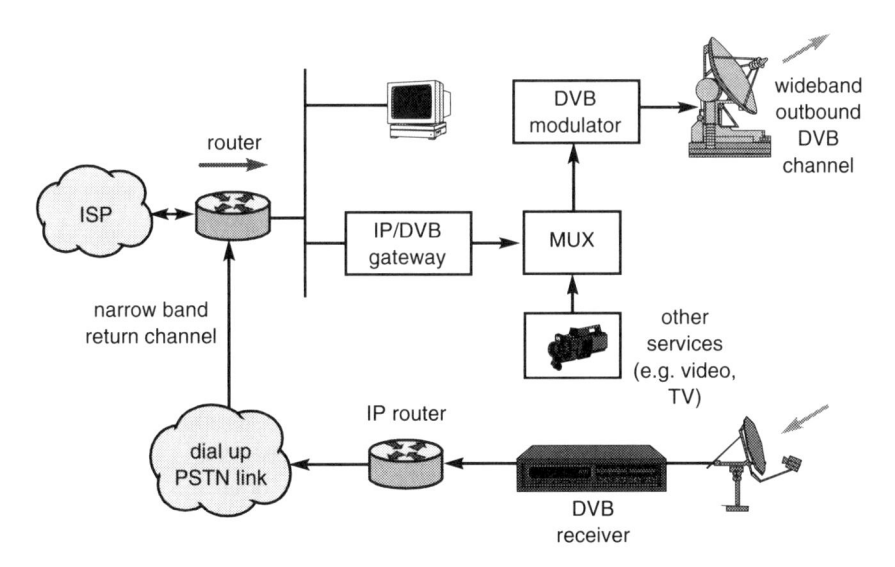

**Fig 10.5** IP via satellite using MPE in DVB.

The IP-DVB gateway, a key component, encapsulates the IP datagrams into MPEG transport stream packets and generates a transport stream, which is then fed to a DVB multiplexer. This stream can be combined with other data or audio/video streams in order to be transmitted as a single carrier via satellite.

Users can be equipped with either a multimedia PC (including a plug-in DVB receiver card and a PSTN modem) or combined integrated receiver decoder (IRD)/ IP router depending on the requirement for a LAN connection. The total retail price of a DVB receiver card and receiving antenna kit is around £200. The retail price of an integrated receiver and router is currently around £300 to £600.

The process for establishing Internet access is as follows. A dial-up connection is established via the service provider to the Internet. Once the user has been authenticated and the connection has been set up, user requests are transmitted over the PSTN line and on to the Internet whereas any return traffic is routed to the satellite uplink site.

Satellite IP over DVB including Internet access, with a terrestrial return link over PSTN/ISDN, was successfully demonstrated in a recent trial, named Convergence 1, in which BT participated. Typical file transfer rates of 1 Mbit/s were demonstrated using FTP.

Presently these systems are operated in isolation to terrestrial broadband Internet services, such as ADSL. However, ADSL may not be available to a significant fraction of the UK population for the short-to-medium term future. This is either because the distance between the exchange and the user is too great for current ADSL technology or because the exchange will not be fitted with ADSL equipment. Satellites, with the key attributes of wide-area coverage and fast flexible access, could be used to complement broadband terrestrial delivery for certain services and this calls for a degree of integration.

## 10.4 System Integration

With the drive towards an 'anywhere, any time' communication philosophy, the need for true system integration will become paramount. An example would be the ability for a user to roam seamlessly between satellite, terrestrial and mobile networks. However, before this can be achieved, common interfaces and signalling protocols will need to be defined. At the interface points the issue of differing technical characteristics between network types will also need to be addressed, e.g. the difference in latency between satellite and terrestrial networks.

One key issue is where to locate the prime functionality of an access network — whether it should be at the edge near the interface to the core network or be distributed across the access network. 'Prime functionality' refers to the ability to intelligently route multi-service (IP, ATM, frame relay, etc) traffic across a broadband access network. An example would be the capability to dynamically reroute circuits by virtual path and virtual switching, including the ability to specify different levels of traffic shaping and management. In the terrestrial environment it is commonplace for ATM switches and IP routers to carry out such processing above the physical layer, i.e. at the network and transmission layers. In contrast, satellite networks are predominantly transparent to layers 2 and above.

*An Overview of Satellite Access Networks* 165

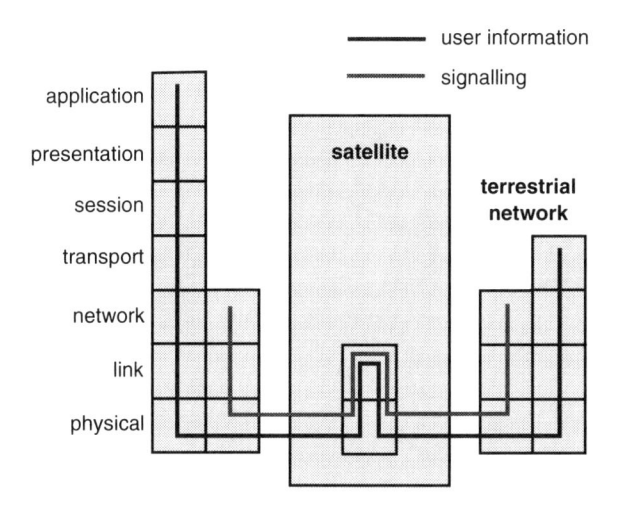

**Fig 10.6** Satellite access network functionality.

However, this scenario will change with the introduction of future broadband satellite systems that use higher layer on-board processing (OBP). This is shown schematically in Fig 10.6.

There are a number of issues relating to OBP. An advantage of OBP is reduced signalling overhead and latency achieved through inter-spot beam switching which avoids having to relay the traffic back to a land earth station (LES) which would be necessary in the case of transparent satellites. At layer 1, this directly translates to more efficient use of the radio spectrum.

The disadvantages are risk and cost of putting key intelligence on board a satellite, additional mass and power consumption, and the fact that business cases demand satellite lifetimes of at least 12 years, whereas the typical lifetime of a terrestrial ATM switch is 5 years.

Another key issue relates to the integration of present and future satellite access networks into the terrestrial and mobile domains using defined interfaces and architectures. These defined interfaces and architectures are required to address issues right across the OSI model. The issues range from the routing and switching that is required at the IP and ATM layer, including signalling convergence, to what level of traffic shaping and connection admission control is required and who controls the network GoS/QoS.

A very high level architecture for global networking capability is shown in Fig 10.7.

One way to address these issues is to review what synergies there are between the various terrestrial broadband standards and how these can be mapped on to the future broadband satellite proposals. After reviewing the predominant terrestrial

## 166  System Integration

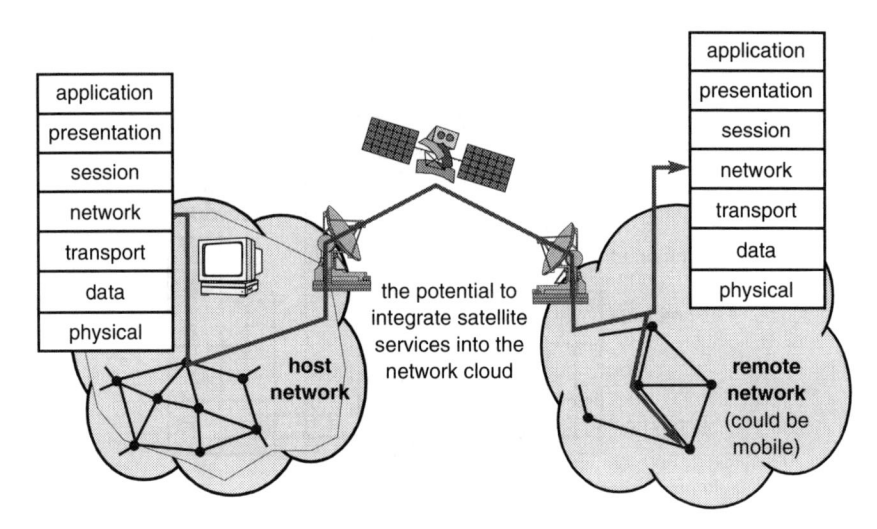

**Fig 10.7**  Global networking capability.

standards and recommendations (namely FSAN, ATM Forum, DAVIC and UMTS), it is clear that the following similarities exist:

- the use of the ETSI/ITU-defined VB5 architecture at the service node interface (SNI);
- the use of the ATM Forum user/network interface, network/network interface and private network/network interface architectures at the user node interface (UNI);
- the use of the ITU-defined Telecommunications Management Network (TMN) based manager-to-manager communications for end-to-end service management;
- for service provision, maintenance and cessation, the use of real-time management and control, broadband bearer control channel and Q.2931 signalling protocols;
- the common use of physical interfaces and service sets;
- the common use of asymmetric and symmetrical services.

Figure 10.8 shows a mapping of the various access standards models, FSAN, ATM Forum, DAVIC and UMTS to the satellite access case.

Hence it is recommended that future satellite access networks should consider the adoption of such interfaces to facilitate the seamless integration with terrestrial and mobile networks. This work is currently being progressed through ETSI SES.

Having defined the architectural interfaces of a satellite access network, the next stage is to define the management architecture required to control the services running over the network. This is another area where there are major differences

## An Overview of Satellite Access Networks 167

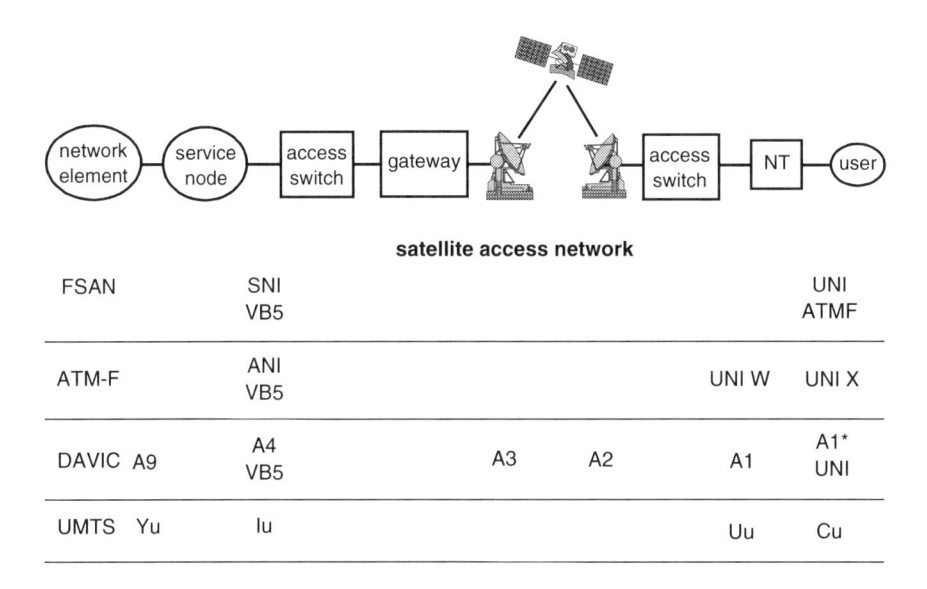

**Fig 10.8** Standards mapping to satellite access.

|  |  |  |  |  |  |  |
|---|---|---|---|---|---|---|
| FSAN | SNI<br>VB5 |  |  |  |  | UNI<br>ATMF |
| ATM-F | ANI<br>VB5 |  |  |  | UNI W | UNI X |
| DAVIC A9 | A4<br>VB5 | A3 | A2 |  | A1 | A1*<br>UNI |
| UMTS Yu | Iu |  |  |  | Uu | Cu |

between present day terrestrial and satellite networks. For example in the terrestrial environment, end-to-end management can be achieved using protocols, such as the simple network management protocol (SNMP) and the common management interface protocol (CMIP) at the network management layer, and TMN, CMIP, CORBA and proprietary protocols at the service management layer.

In contrast, generally, in the satellite environment, only proprietary management architectures are used to date with no dynamic interworking between satellite operator and terrestrial network operator. For example, presently most space segment bookings are done manually via paper or voice transactions.

This is greatly at variance with the vision for the future whereby a user should be able to seamlessly roam between satellite, terrestrial and mobile networks with the capability to request dynamic service features, for example video-on-demand capabilities. The vision of integrated network management is shown in Fig 10.9.

This will require detailed management and signalling traffic between the various network elements right through from the content provider to the end user. Hence future satellite access networks should aim to offer standard management and signalling interfaces for the purpose of interworking. The general case, which is more complicated, is where the satellite operator is a different organisation to the operators of either the originating network or the remote network. Figure 10.10 shows the interfaces necessary for this general case.

Trials of network management protocols such as SNMP and CMIP at Adastral Park indicate that there are no major problems when operated over satellite. SNMP has limitations, such as its mandated use of UDP/IP, which makes it inherently

168  System Integration

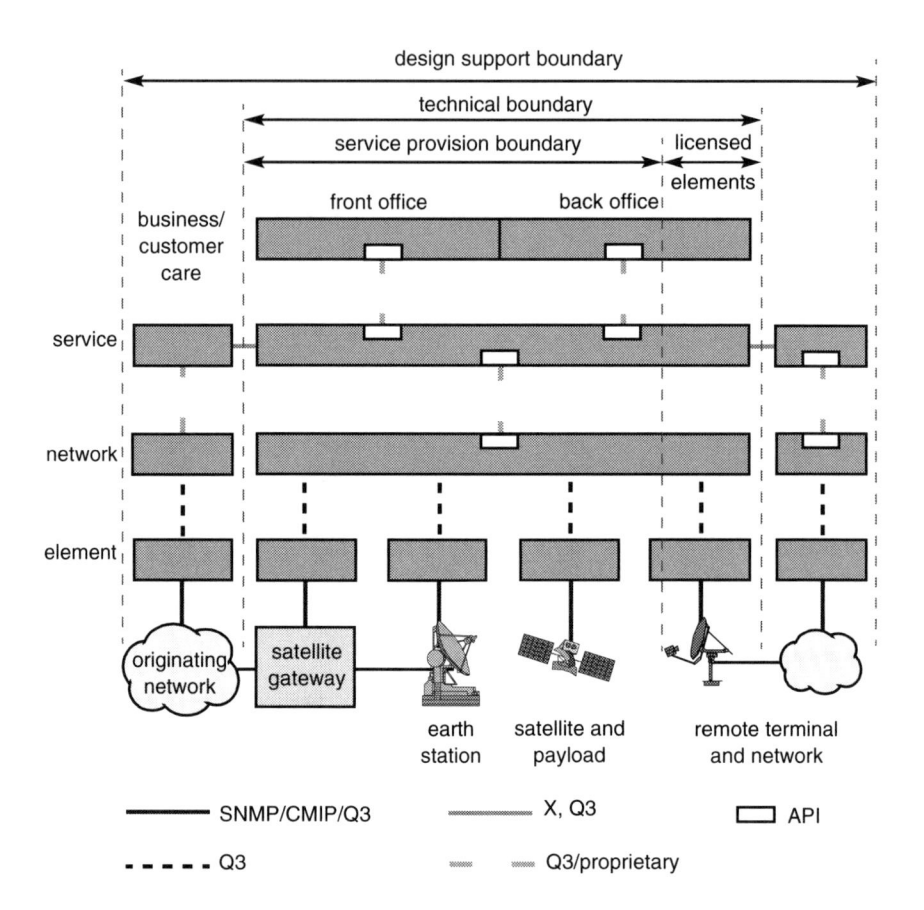

**Fig 10.9**  Integrated network management architecture.

unreliable, and the fact that it relies on polling and traps means that it takes a long time to download management information base (MIB) values, if there is much data to transfer. CMIP is much more powerful and uses TCP/IP — however, it also carries more overheads. CMIP is more suited to some network management functions particularly if the management traffic has to go over satellite links.

Some other issues raised by the subject of interworking, which are the subject of on-going work at Adastral Park and within the standards bodies, are given below.

- Who manages the end-user equipment — the satellite operator, the network operator or the service provider and how, using SNMP, CMIP or CORBA?
- What management interaction is there between the end user, satellite operator and the network operator for the provision of additional bandwidth on demand, e.g. if a new user joins a videoconferencing session?
- Where should the management boundaries be?

An Overview of Satellite Access Networks    169

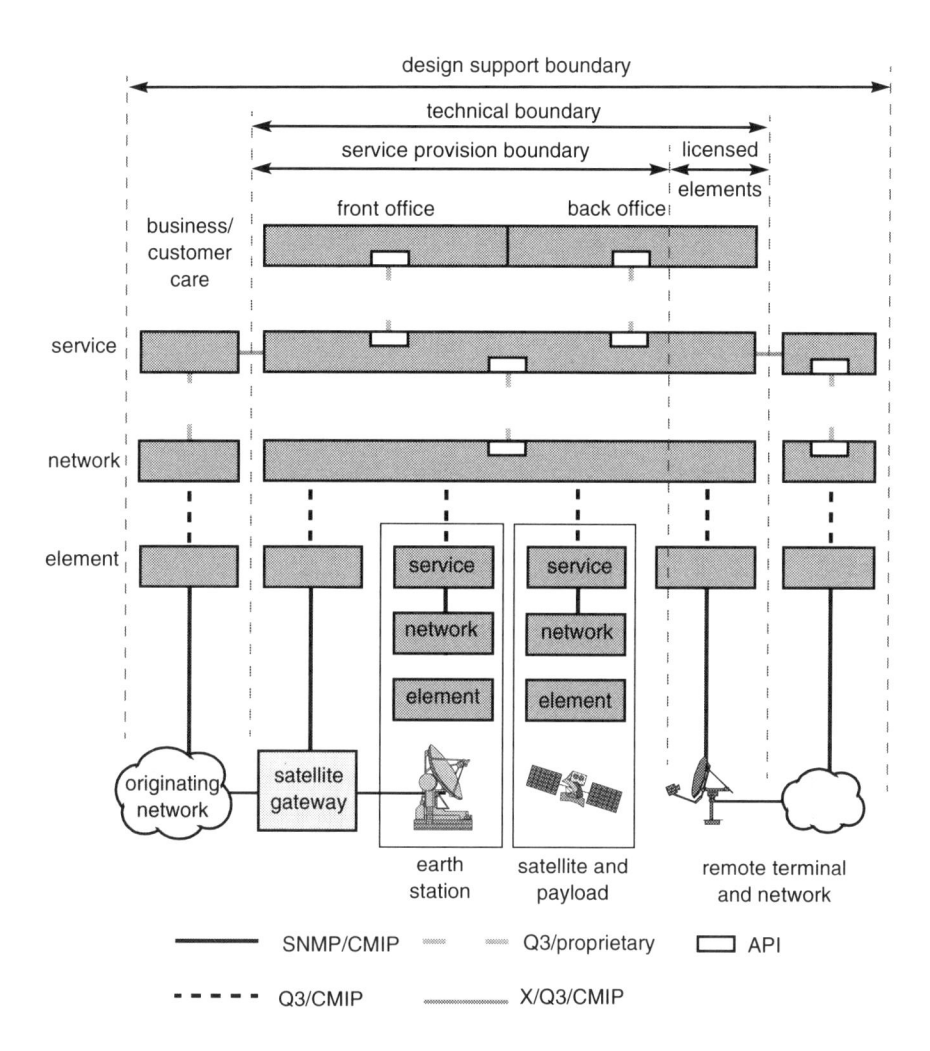

**Fig 10.10**  The general architecture assuming a third-party satellite provider.

## 10.5    Conclusions

Satellites are being used increasingly for IP delivery in the access network. Their main attribute of wide-area coverage allows fast and flexible deployment. They are used currently for ISP backbone connection and for delivery of IP services to users over a variety of link-layer protocols. The most suitable link layer for use with low-cost mass-market receivers is DVB and several examples exist of Internet satellite access systems using this technology.

Future satellite systems will be even more suited to the access network, incorporating layer-2 switching and narrow spot beams, enabling even more bandwidth to be delivered to the user at low cost.

The time frame for the next generation systems is 2003—2007. Integration is a key issue to efficient use of satellite resources, both at the user traffic level, the signalling level and at the network management level.

# 11

# OPERATIONS FOR THE IP ENVIRONMENT — WHEN THE INTERNET BECAME A SERIOUS BUSINESS

## J Ozdural

## 11.1 Introduction

This chapter describes the role of operations and the typical environment found at an IP operations centre. One of the key factors in satisfying customers is the quality of the service provided by the operations team. Two key ingredients support good operations — people and tools.

BT has a long history of operating IP networks — its Internet operation was established in 1994 when its public Internet service was launched; however, it has been operating private IP networks for over 20 years. Over this time BT has gained valuable experience of operating IP networks in a rapidly evolving environment and has optimised its operations for best performance. The chapter captures the experiences gained and systems developed in the field of Internet operations. The most important factors in operating IP networks are:

- the organisation of operations;
- a clear understanding of the roles and tasks;
- a structuring of customer support into tiers or lines so that best use is made of people with valuable IP skills and customers receive best service;
- the development and deployment of suitable tools;
- a clear definition of the service surround required to maintain an operations centre.

## 11.2 Operations Organisation

There are basically three distinct units within an operations function from a top level. The first is customer reception, then first-line diagnostics/second-line specialists, and finally technical services. Figure 11.1 indicates the typical flow of tasks and data into an operations unit. The structure described has been implemented within the BT IOC and has proved to be highly successful at maintaining and improving operations integrity.

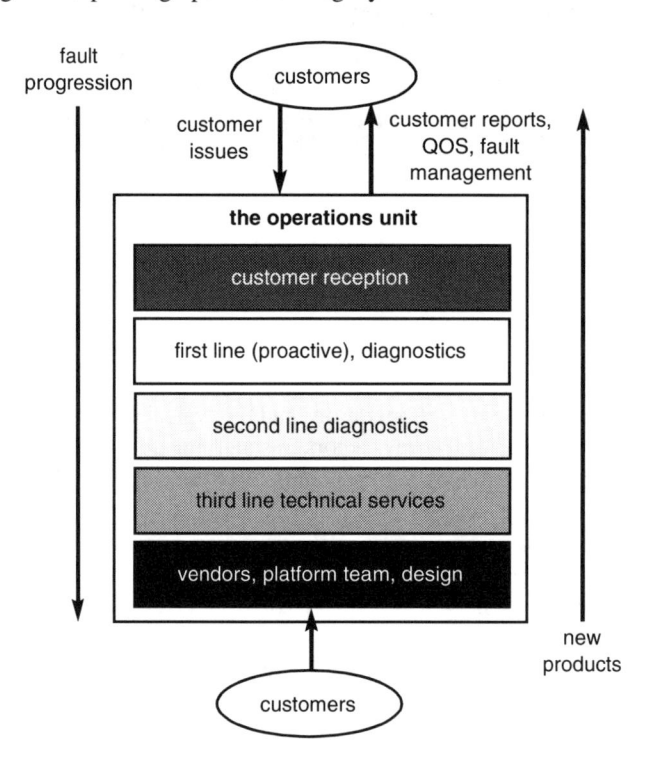

**Fig 11.1** Operations organisation.

In Fig 11.1, the first-line and second-line are shown as distinct units. This is only true in terms of the complexity of tasks performed by the units. In reality, at the BT IOC, both units report to the operations manager. New products are those which are innovation or market driven and as such the in-life support for these products needs to be established. This role and the ownership of complex problems are the province of the technical services (TS) team. The TS team members consist of the most experienced network and commercial people in operations. They will typically have around 10 years relevant experience each. This does not mean that the first- and second-line operations units would not have experienced people in them, rather that the concentration within TS is higher. New graduates and experienced operations managers would populate the first-line operations unit role, whereas the second-line teams would comprise more specialised product champions, for instance WAN specialists and dial operations. The basic policy is strength in depth — success is built upon the depth of commercial and technical expertise available.

## 11.3 Functions that Need to be Performed by Operations

The following list of operational roles and tasks is not exhaustive and is presented to give a flavour of the myriad of interdependent features that comprise the task of network operations.

- Network management:

    — to maintain service on a 24-hour basis, which requires providing continuous engineering cover (operations and technical services) and duty manager cover;

    — to ensure network availability and minimise outages;

    — to be proactive in the management of customer installations;

    — to use a network management system (NMS), such as OpenView or Omnibus;

    — to maintain a secure access method and security policy database (SPD);

    — to fine-tune and change the network via strict planned engineering works (PEW) guidelines;

    — to maintain the relationships with and alliances between other Internet service providers (ISPs);

    — to maintain service level agreements (SLAs) and memoranda of understanding (MOUs) with third party vendors such as asynchronous transfer mode (ATM), synchronous digital hierarchy (SDH) and leased circuit providers to ensure that contracts are within specified targets.

- Configuration management:

    — to keep and update the configuration of all elements that make up the operation;

    — to maintain network diagrams and core network connectivity diagrams;

    — to review and audit the network components.

- Capacity management:

    — to capacity manage and plan for network growth directly and via third party planners;

    — to provide proactive management of the core network infrastructure capacity.

174  *Functions that Need to be Performed by Operations*

- Customer service management:

    — to provide clear and concise access points, i.e. Web-based to disseminate customer information and to report faults;

    — to publish service availability and QoS (quality of service) targets against actual results (QoS targets are service issues such as minimum time to restore customer connectivity, availability of service (over 24 hour period, weekly and monthly), PCA (percentage calls answered), customer satisfaction surveys — in addition, technical QoS indicators are available such as packet loss and delay (delay to reach customer sites and per packet));

    — to maintain and improve on QoS target indicators;

    — to provide customer ownership and feedback of faults and issues;

    — to provide proactive and reactive management of customers and infrastructure;

    — to be aware in detail of product and service variations;

    — to maintain a customer information database.

- People management:

    — to provide training and keep up to date with the latest technologies and services;

    — to provide a safe working environment in conjunction with health and safety regulations;

    — to maintain and develop qualified expertise to support the network;

    — to develop and improve its people in conjunction with IiP (Investors in People);

    — the Investors in People Standard provides a framework for improving business performance competitiveness, encouraging excellence in the field of human resource development and creating a culture of continuous improvement (see Appendix for more details).

- Service development:

    — to accept into service (AIS) new products and services;

    — to 'shift left' experience by the use of automation and OSS (operational support systems) tools (the term 'shift left' in this context is used to describe a set of plans and procedures that allow experience and training from highly skilled engineers to be imparted to front-line operations teams — this is not to say that front-line operations are less skilled, only less focused).

## 11.4 The Roles

This section describes the major roles performed by people and systems at the BT Internet Operations Centre in more detail.

### 11.4.1 Customer reception

Customer reception manages the customer experience. It is the first point of contact for the reporting of customer faults and issues. The team provides 24-hour reception for customer faults and queries via the Internet, e-mail, fax and telephony. In the BT IOC, customer reception will own customer issues and the progress and tracking of them until completion. The members of this team need in-depth customer skills to acquire and address customer issues. In addition, they require automated tools to allow them to provide technical feedback to the first- and second-line operations teams. This is done via a structured questioning approach built on engineering experience. An essential element of customer reception is the maintenance of a full and comprehensive customer database that holds contacts, circuit details, configurations, options and fault histories. The sub-databases should be accessible via a relational database that allows simple, complex or precompiled queries. The customer reception team maintains and produces statistics on QoS and availability to customers on demand and at specified periods. The customer reception team has agreements with product and sales teams to provide them with performance and customer satisfaction information. The essence of customer reception is a speedy, customer-focused approach coupled with dedicated ownership of customer issues. That is a one-stop shop to customer issues. If the fault is not of a technically challenging nature, e.g. a circuit fault, the customer reception team would own the task of restoring service. The customer reception team maintains an escalation path to enable the restoration of service or resolution of a customer issue, i.e. they maintain a list of management contacts, for services supplied from within BT and by third party suppliers, so that, in the event of a failure to resolve a customer problem in time, the problem can be escalated to an appropriate management level for resolution. The customer reception would also provide the customer with planned engineering works notifications and gain customer agreement for the planned outages. Customer issues are managed via the trouble ticket system. A unique ticket is created on the receipt of a customer issue. The ticket contains the customer details plus the details of the customer issue. The progress or otherwise of the customer issue is detailed in the ticket system. The ticket system will have the ability to produce statistics of resolve times and to what stage the ticket has progressed. This system is a key indicator to feed back to the customer, and to all teams in the operation, on the effectiveness of the operation.

## 11.4.2 First-Line and Second-Line Diagnostics (Real-Time Operations)

These teams maintain a 24-hour rota to address customer issues. The fault or query (still owned by customer reception), if not of a simple nature covered by customer reception tools and knowledge, would be passed to the first- and second-line diagnostics teams. The first-line team is responsible for the proactive and reactive management of the core and customer premises equipment (CPE). This duty would be responsible for using more in-depth tools for the resolution of customer issues. The tools in question, used at the BT IOC, are OpenView (SNMP — simple network management traps and alarms) and Omnibus (alarm element manager). These tools, and other BT-developed in-house NMS applications, are monitored by this team. The tools report on device availability, link and core capacity, security, and usage; they also proactively and reactively manage the network and CPE elements. Network management is achieved by receipt of traps and alarms of network events such as link-up/down, intrusion events and topology changes. The NMS systems in addition actively poll the network elements and CPE to inquire on their status. On a variation from the QoS or the receipt of an alarm indication from the NMS, a fault ticket is created and the variation investigated. The use of a set of diagnostic tools, in addition to engineering intelligence database systems (EIDSs — history fault resolution techniques), is used to determine the nature of the fault or query. If the diagnostic test reveals a previous record of the instance with a known resolution the first-line team continues to deal with the fault. If the first-line team cannot deal with the issue within a defined period or if the issue is beyond their experience, the second-line team (product specialists) are called in.

The second-line team consists of highly experienced product specialists trained in the main product areas. At this point the tools are the same as the first-line team — it is the engineering knowledge that is much greater. Within the defined QoS targets, the second-line diagnostics team will attempt to deal with the issue. If for instance this issue were a known bug or hardware fault, the ticket would require no further escalation via the on-call duty manager. The maintenance of operations management and integrity is the responsibility of the on-call duty manager, usually taken from the ranks of the first- and second-line diagnostics team. The on-call duty manager is responsible for the escalation and resolution of customer issues. In addition to QoS issues (such as time expired issues or problems outside the technical experience of the first-line team) the second-line operations team receives new products and services delivered via the third-line technical services team. The third-line team specifies or creates support frameworks and OSS tools in co-operation with the second-line teams who accept them into service.

The second line also have managers that operate the PEW process for planned outages for network improvement.

### 11.4.3 Third-Line Technical Services (Elapsed-Time Operations)

This team has many roles, which include among others escalations from second-line operations for faults that are approaching critical status in terms of time expiring and beyond second-line operations remit and experience. The TS team members are product champions and own the delivery and pilot support of products through the AIS process. The team operates a policy to move complex tasks from highly skilled technical services into customer reception and first line by the use of tools and processes. This policy is traditionally referred to as a 'shift left' policy, i.e. to move OSS and tools to customer reception and first-line diagnostics. The shift left policy is also attained by TS conducting in-house training of a technical and commercial nature.

This team has a high concentration of the most experienced and talented network engineers. They can converse with the IT directors of customers and competitors. They also participate in the network design of new and existing products. The TS team provides the primary interface to the vendors and to the design authorities. The interface role is important with regard to faults and bugs that require a high level of technical contact with vendors and design teams. Due to this level of contact some industry recognition, i.e. vendor-accepted certification, is desirable for these individuals. The TS team maintains and runs the operations SPD for server and network elements. They actively groom the network and handle issues of network entropy. They provide consultancy and value added services (VASs) (see section 4.4). At the BT IOC, the team is split into three distinct units — dial platform services (to cover the explosive growth in dial products such as BTClick and BTInternet), BTnet platform specialists that deal with WAN and policy and technical issues on the infrastructure and agreements with co-operating Internet service providers (ISPs), and thirdly 'plan and provide', who look at all capacity issues on the network.

### 11.4.4 Consultancy

Network experts with several years' real-time experience are required to provide consultancy services. Examples of these consultancy services are:

- performance analysis and feedback to customers;
- network design services for customers;
- integration, expansion and migration of networks;
- end-to-end service design and integration;
- bespoke application delivery;
- network security and integrity investigations;

- special fault investigations (SFIs) and event diagnosis of customer and core networks;
- end-to-end service reviews.

A related organisation to TS is capacity and infrastructure (plan-and-provide) management. The main role of this organisation is to maintain the network inventory and location, and configurations of the live equipment. The capacity team also maintains the connectivity diagrams of the network and are also responsible for monitoring the use of links and infrastructure and specifying the growth of that infrastructure.

## 11.5 Tools Used

Due to the rapidly evolving nature of the IP business some of the network management system and operational system support tools mentioned in this chapter are being continuously developed. BT's ability to integrate best of breed tools developed by third parties, supplemented by tools developed in-house by BT, is a major contributor to the efficiency and effectiveness of its Internet operation centres.

The tools shown below are the basic OSS building blocks for maintaining a network operations service.

### 11.5.1 Customer Inventory Database with Search Engines

This database comprises information about location, contact, service description fields, customer configuration and services. These tools accept complex, precompiled and simple macro queries. An example would be to list all or a particular customer site(s) plus fault history. At the IOC the system initially consisted of several bespoke systems that have been integrated into a single customer database system. The integration has had several releases over the years, but that is due to increased functionality and is a demonstration of the future proofing capability of the customer inventory database.

### 11.5.2 Customer Reception Front-End Tools

Automated call distribution (ACD) systems linked to customers' details with PSTN QoS reception indicators (in this scenario the QoS indicators are of the form of PCA and elapsed time to resolve customer issues, among others) interrogate an incoming voice call to automatically bring up customer details. If the ticket is received via the Internet, the ticket system is intelligent enough to attach the customer details based on Internet address. This Internet-based system is linked to a simple connectivity and diagnostics test tool that pastes data directly into the customer ticket. This would form the basis of a ticket escalation system linked to QoS indicators. These systems consist of 'off-the-shelf' packages and standard BT telephony products plus ACD call-centre products [1] integrated by BT.

### 11.5.3  Customer Usage and QoS Automated Tools

These tools provide up-time and usage information of the customer's connection to the service provider, including up-time, outages, utilisation, and CPE information. These systems consist of an entire range of QoS indicators, some 'off-the-shelf', but mostly developed by BT. Where the systems are bought in, they have been integrated into the alarm-handling systems operated by BT. This means that each bought-in system has been 'enhanced' for use in the BT IOC environment.

### 11.5.4  Polling Diagnostics for Customer Reception and First-Line Diagnostics

Automated front-end Internet Web-based tools check:

- ping (are you there?);
- poll (how are you feeling?);
- reverse DNS (what is your name?);
- IP addresses;
- ingress and egress points of traffic from a customer view and from the Internet;
- implants, such as Wandel & Goltermann Domino, to test and monitor network and customer traffic flow;
- difference engines to spot changes in customer and network configurations.

Although it is possible to use bought-in equipment, the majority of the equipment and systems used have been developed in-house (Web-based, Oracle-based database extraction tools for instance), using the engineering experience gained from the IP operations environment.

### 11.5.5  Structured Questioning Tool

This tool gathers quality customer information on a customer issue. This is an intelligent knowledge-based system (IKBS) that builds customer and fault analysis questions based upon received 'expert' input. An IKBS system is one that has the ability to learn based on the expertise programmed into it. The system takes

180  *Tools Used*

scenarios, such as a circuit fault, and applies various solutions based on the real experience of solved tickets. As the data and the methodology is only as good as the 'experts' that program the system, it will improve with time. The tool is useful for customer reception to pass on diagnostics to first-line diagnostics teams. This tool consists of real off-the-shelf products developed for use and sale by BT Syncordia and employs structured questioning techniques and processes developed by groups such as BT MCS (major customer services). The BT MCS group operates at the BT IOC in the role of customer reception.

### 11.5.6  On-Line Fault Reporting Tool and Customer Diagnostic Tool

These tools give customers on-line access to reports and the ability to view the progress of their ticket or query. The customer diagnostics tool allows customers to test their own end systems with a limited set of tools used by customer reception. This allows the customer to view their connectivity from the Internet. The reports are available to both BT Syncordia and BT IOC customers. Initially these reports were derived from bespoke developed applications. These have now been integrated into BT industrial-strength systems. Syncordia currently offers these self-test facilities.

### 11.5.7  NMS Systems to Receive Traps and Alarms and Raise Automated Fault Tickets

This tool for instance raises alarms on OpenView, which are interpreted by the higher element managers such as Omnibus (from Micromuse). It is a basic building block designed to receive SNMP, poll and log information from different vendor equipment. The output raises red, amber and green (RAG) alarms on the basis of network events at layer 2 (links and backbone) and layer 3 (routers). Although Omnibus and OpenView are off-the-shelf products, BT IOC and Syncordia have added value to the systems by writing several additional alarm modules and alarm database interrogation systems.

### 11.5.8  Security Intrusion, Modelling and Anti-Hacking Tools

These tools are used proactively and reactively to test the service provider and customer network. They exist inside and outside BT's network service boundary to give BT the ability to test their SPD. Within BT, these systems comprise off-the-shelf products plus specific in-house procedures and tools. Firewalls and security implants at network points monitor service SPDs. War-gaming implants at remote sites test front-end network security; this feeds into a back-office repeat-event database.

### 11.5.9 Capacity Management Tool

This tool has been designed to see if a particular point of presence (PoP) or customer network is working within specification, e.g. whether there are enough ISDN terminations present. This tool is part of the planning function, and has been exclusively developed by BT.

### 11.5.10 Network Capacity Management Tool and Alarm Indicator

This tool is similar to that described in section 5.9, but actively monitors the network for capacity issues. For instance a new PoP access point is required as a particular site's access passes a certain threshold. This threshold could be set to 80% and an alarm on the NMS created to request an increase in PoP capacity. Major tools have been developed in this area that integrate the PSTN statistics with the IP network usage. These tools have all been designed and developed by BT, based on real engineering experience.

### 11.5.11 Web-Based Access to the Majority of Access and Polling Tools

These are used particularly in the dial platform management space and have been developed by BT.

### 11.5.12 Modelling Tools for 'What If' Scenarios

These tools are designed, for instance, for determining the impact of a new service or of new network equipment on an existing service. A combination of in-house and off-the-shelf tools are used by BT's Adastral Park community to test and develop the services.

### 11.5.13 EIDS for the Analysis of Repeat Faults

The intention is to make this an intelligent knowledge based system to give guidance to operations in fault-finding histories. This is a bolt-on to the Omnibus system with value added by BT systems development.

### 11.5.14 Centralised NMS Infrastructure

Tools are provided to poll and check the consistency of both the live data, and that held on the fall-back sites.

### 11.5.15 Trap Event Collection (Link Up and Down Messages to Active Database QoS Engines)

A historical back-end engine looks at alarm history which feeds into QoS reports available on the Internet to customers. The trap event collection produces proactive reports based on unusual link activity, i.e. flapping links that would indicate circuit problems after a number of instances. The trap event collection tool automatically generates a trouble ticket to the carrier network to request a test and feedbacks to operations after running a customer reception diagnostics tool to gather more information. These systems are in place as bespoke applications at the BT IOC and are being migrated into industrial-strength applications by BT.

### 11.5.16 Unique Polling Engines Backed by Diagnostic Tools

The polling engines have the ability to view the network object both from inside the ISP infrastructure and from outside (secure access). These systems have been developed for use in BT as bespoke applications and are being migrated to industrial-strength applications platforms.

### 11.5.17 Automated Router Configurations

These provide modelling, security and resilient access, and management reporting. BT IOC, Syncordia Solutions and Concert [2] have these systems in place using bespoke and of-the-shelf items.

### 11.5.18 Event History, Log and Trend Analysis in Conjunction with Audit Reports

This tool provides the number of faults reported by the customer and by proactive tools, and also the availability of the customer network, and usage of the customer networks. This trend analysis tool is used by the capacity management team and by products and sales. The capacity management team uses the information in conjunction with infrastructure reports to increase the network capacity in conjunction with demand. The products and sales team uses the information to sell more bandwidth and connections based on customer usage. BT has developed these systems and they are used throughout the business units.

### 11.5.19 Connectivity Checking Tool for Customer Ingress and Egress Points

This tool is available in two forms — one that the customer can view securely from the Internet and the other that the operations and technical services teams use. The basic function of the tool is to be a 'looking glass' from inside and outside the network to observe the source/destination routes used to and from the customer network, plus the actual route taken. The reason for the two variants, one for customer and one for operations, is that the operations variant has access to the internal ISP carrier/router network with full engineering data.

### 11.5.20 Archive and Back-up of Routers and Configuration Tools

These tools run at set and random intervals to observe and back up the configurations of customer and ISP networks. The ISP network is that network which supplies service to the customer, for instance a PoP. Routers and data networks become more complex by the day and the number of configuration variants available multiply at the same rate. Therefore to configure a customer connection effectively, a list configuration tool must be available to create a configuration based on simple questions and answers. Service delivery will tend to use the configuration tool, while operations will use the archive and back-up tool. However all these tools are linked by an over-arching configuration tool. The reason for this is that, as soon as a service delivery event is completed (adding a customer connection), the routers and connection are then eligible to move into the archive and back-up phase without user intervention. BT has exclusively developed these tools by using off-the-shelf items and in-house innovation.

### 11.5.21 Restoration of Configuration Tool

This tool holds several iterations of customer and infrastructure configurations in addition to hardware that makes up the connection. If $N$ is the current iteration of the customer network, an on-line $N$, $N-1$ and $N-2$ version is also available, i.e. the current version, the last modified version and the version before that. In addition, off-line tools hold compressed iterations and configurations going back six months. This tool feeds into the customer and performance reports for the customer and the ISP network. These tools have been developed by using off-the-shelf items and BT innovation.

### 11.5.22 Operating System Upgrade Tool

This tool allows the upgrade of a customer router or the ISP infrastructure based on field alert notices or triggers from technical services. The field alert notices are reports of instances of behaviour that require a change to the router's operating system and parameters. The tool maintains the current operating system on-line and associates it with the customer connection. The tool works in conjunction with the archive tool to maintain the configuration and can download at predefined times. The tool can also handle multiple upgrade instances. BT has developed these tools using off-the-shelf items and its own in-house innovation.

### 11.5.23 Password and Access Management Tools

This set of tools performs a number of functions:

- password resets activated at time intervals and at the request of the customer — this applies to the situation where the customer owns their network connectivity and uses the ISP to maintain the configuration (in this instance the customer may maintain the configuration but devolve password management access to the ISP);
- in all other cases password resets are performed for and on behalf of the operations centre — this is for the maintenance and support of correct levels of secure access for operations people where some engineers may require access to all systems (access all areas (AAA)) and some may require access to a subset of systems and features;
- in both the above cases, the authentication mechanism requires that a set of systems be set up to validate the requests — one of the systems used is 'public key infrastructure' (PKI) which uses encryption to validate and identify genuine requests from the customer or ISP engineers;
- management reports can be produced on the patterns of access to look for unusual activity and excessive number of reset requests — these reports are used to add to the secure management of the operations environment.

### 11.5.24 Configuration Management History Reports in Conjunction with Performance Management

As indicated, this tool is a feature that is added to the archive and back-up tools to give configuration history in the customer or ISP network.

### 11.5.25 Disaster Recovery Procedure Tool

This tool holds the hardware inventory of the network and is linked to network diagram tools. In addition, this tool is linked to the archive and configuration database.

The tool, in the event of a failure of a customer or ISP network connection, will bring up alternate paths to restore service. Where the failure is fatal, i.e. dead hardware, the tool will arrange/prepare for the despatch of a replacement unit in conjunction with operations PEW procedures. These tools have been developed by BT using off-the-shelf items and in-house innovation.

## 11.6  Service Surround

Figure 11.2 shows the basic attributes that are required for the operational 'service surround', i.e. BT's commitment to customer service.

These attributes act as anchor points to hold together the fabric of the operation. They are, in effect, the values that would feed into an IiP programme (see the Appendix) and into individual annual performance reviews.

People are a company's best asset. The values (see Fig 11.2) indicate the attributes associated with maintaining and keeping in place highly professional and motivated people. They are about valuing the people and the reputation of the operation, education (customer and technical) to vendor certification levels, assessments and programmes.

The programmes consist of quality management systems to make the job easier and more consistent, and improve customer service. Pay and remuneration are both important factors — the total package offered needs to be in line with market rates to maintain low people turnover. Turnover is an important factor in the success of any operation, as those individuals that have acquired the skills required need to be retained to bring on more junior members of the team. Strength in depth is also important.

Service commitment by use of customer account and QoS management, restoration procedures and best practice models are all attributes that show the measure of the commitment of the operation to its customers and shareholders. By the use of alliances, best-practice models (off-the-shelf designs and products), network resilience (elimination of single unit failures) and service QoS guarantees (publishing of QoS service targets, such as PCA, to customers), the operation demonstrates its commitment to customer service.

186  *Service Surround*

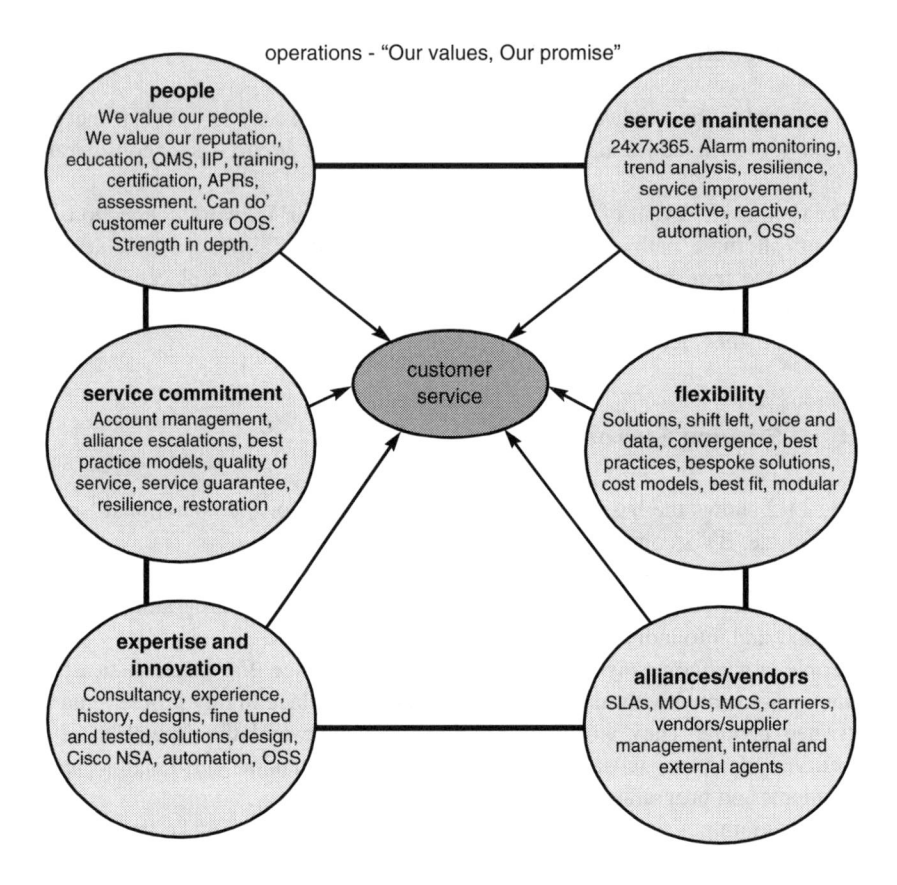

**Fig 11.2** End-to-end service surround.

Expertise and innovation are demonstrated by maintaining expert, experienced personnel who will create OSS tools for automation, through the use of innovation and consultancy to meet and exceed customer requirements.

QoS and network availability indicators are used to measure service maintenance, and this attribute, in conjunction with service improvement initiatives, help to improve performance and reduce operating costs.

For flexibility, the operation will use best practice and modular design and, where appropriate, off-the-shelf applications to support the operation. The operation does not preclude any suggestion that enables the operation to gain financial and competitive advantage. This flexibility applies to all operations personnel. The 'shift left' policy is to develop OSS and expertise from third-line to first-line operations. This means that third-line operations would put in place training and OSS for the benefit of first-line teams.

The operation will maintain and exceed QoS targets by use of alliances, such as SLAs and MOUs.

## 11.7 Conclusions

This chapter has shown that the maintenance and running of an Internet operations centre is not a trivial matter and covers a multitude of strategies and disciplines. It is an environment that is unique in bringing together customer management, people management and pure engineering procedures and innovations. It can be seen that strong, established procedures are required as well as direct engineering and managerial innovations, both existing at the same time. People skills and motivation are the key that can make or break an operations culture. A flavour of the tasks and issues involved has been presented. In short, a successful career in operations is very complex and should only be attempted by those with a passion for operations and who are resilient to change.

## Appendix

### Investors in People (IiP)

This initiative is important to ensure that the people who maintain and grow an operation develop themselves, feel valued and contribute to all initiatives at the Internet or network operations centres.

What is the Investors in People Standard? It provides a framework for improving business performance competitiveness. It encourages excellence in the field of human resource development and creates a culture of continuous improvement. The Investors in People Standard draws upon the experience of some of the UK's most successful organisations both large and small. They have found that business performance improved through the development of a planned approach to meet these objectives. The result, in simple terms, is that it ensures that what people can do, and are motivated to do, matches what the organisation requires them to do.

IiP is delivered through the national network of Training and Enterprise Councils (in England and Wales) and the Local Enterprise Councils (in Scotland).

What is IiP about?

- good management practice;
- a quality standard for people.

In an IiP organisation:

- everyone knows about the goals and targets of the business;
- everyone is properly skilled to do the job;

- everyone can explain their job and its importance;
- everyone knows that the business is committed to people development;
- everyone is committed to improvement.

  What are the principles?
- commitment to investing;
- planning, training and development needs;
- taking action to develop necessary skills;
- evaluating the progress towards goals.

  IiP in the context of information services:
- will help to improve the way information services is run and developed;
- is within the 'Investing in our Future' campaign;
- will focus on coaching and development.

## References

1   BT Syncordia Solutions — http://www.syncordia.co.uk/
2   Concert — http://www.concert.com/

# 12

# OPERATIONAL MODELS AND PROCESSES WITHIN BT'S AND CONCERT'S ISP BUSINESS

### J Meaneaux

## 12.1 Introduction

The challenge for an IP network operator is to have flexible processes which allow innovation and development which then allows lightweight processes for new advanced products, but within the same environment to have rigorous and efficient processes for well-established products. This chapter gives an overview of operational models and processes, using BT and Concert as examples. It should help the reader understand how to design and develop operations to be flexible and efficient.

There has been rapid growth in the demand for IP services from the corporate market. Concert and BT products have developed from simple IP access to complex security and eCommerce supply solutions within a short space of time. New IP products and enhancements are being introduced into the corporate market in the BT environment about every three months. How have the operational models and processes changed and developed to accommodate this rapid development cycle? What is the best model to utilise for BT's and Concert's Global IP network service businesses?

## 12.2 What are Operational Models and Processes?

Firstly, it is important to establish exactly what operational models and processes are.

190    *What are Operational Models and Processes?*

### 12.2.1   Operational Model

An operational model exists as a way of defining the business in which we work. It outlines at a high level, without details, the key functions within an organisation in terms of roles and responsibilities, for example, during the delivery and fault-management cycles. Figure 12.1 shows an example of a high-level process design for support from an operational model for the Concert Managed Firewall product. A real operational model will be a complete document that may contain several figures similar to Fig 12.1.

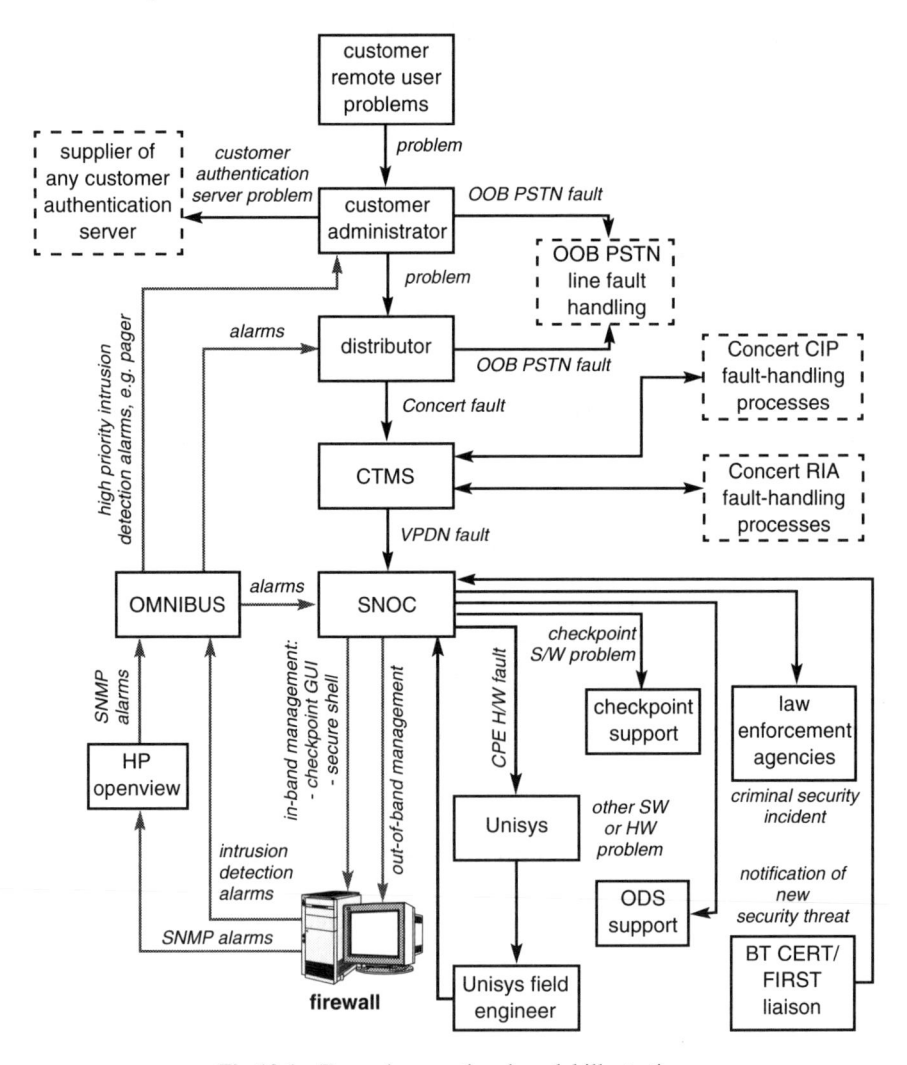

**Fig 12.1**   Example operational model illustration.

*Operational Models and Processes*  191

The model would be largely aimed at the function owners or managers and the model will be developed during the initiation stages of the product life cycle. It will define the systems to be used and the volumes of orders the organisation will have to sustain, usually in terms of volumes of orders and faults per month. With specific relation to a Global ISP, operational models tend to be defined on a product basis due to the need for IP products to go through a more specialised life cycle. Ideally though, the model should be generic across the products to allow optimum use of resources and systems across the portfolio. Operational models are important to the business for several reasons. They can reflect operational strategy and the organisation in general, but, much more importantly, the model goes a long way to indicating the operational costs of support and highlights system development requirements.

### 12.2.2 Processes

Sitting beneath the operational models are the processes, which provide the detail within the functions. Processes will be written every time a new product is released or features added to an existing one. For example, there will be a process for delivering a basic IP service, circuit, router and IP configuration; then, where load-sharing has been introduced as an additional service option, the process will be updated to include the changes and additions to router statements and any additional billing activities. Figure 12.2 shows an example Concert process for order capture and placement for the managed firewall product.

Producing processes has a twofold benefit. They define the work and derived data that provide a large input into systems development. Also, because processes are aimed at the people doing the work, a new person should be able to walk into an organisation, read a process document, and understand what their daily responsibilities will be and what systems they will use to discharge their responsibilities. It is worth pointing out, however, that only the process documents that have been written correctly will enable this.

### 12.2.3 Procedures

Underneath processes is another layer called procedures. These go down to a keystroke level and are particularly useful, for example, for defining order entry guidelines into particular systems. Procedures are generally used where a job demands high accuracy but a lower technical skill set. Procedures are usually produced after a system that supports a process has been developed, and will tell the user exactly what to do within a system. Within the context of a global ISP, generally the higher the technical requirement of the function, the less need there is for procedures.

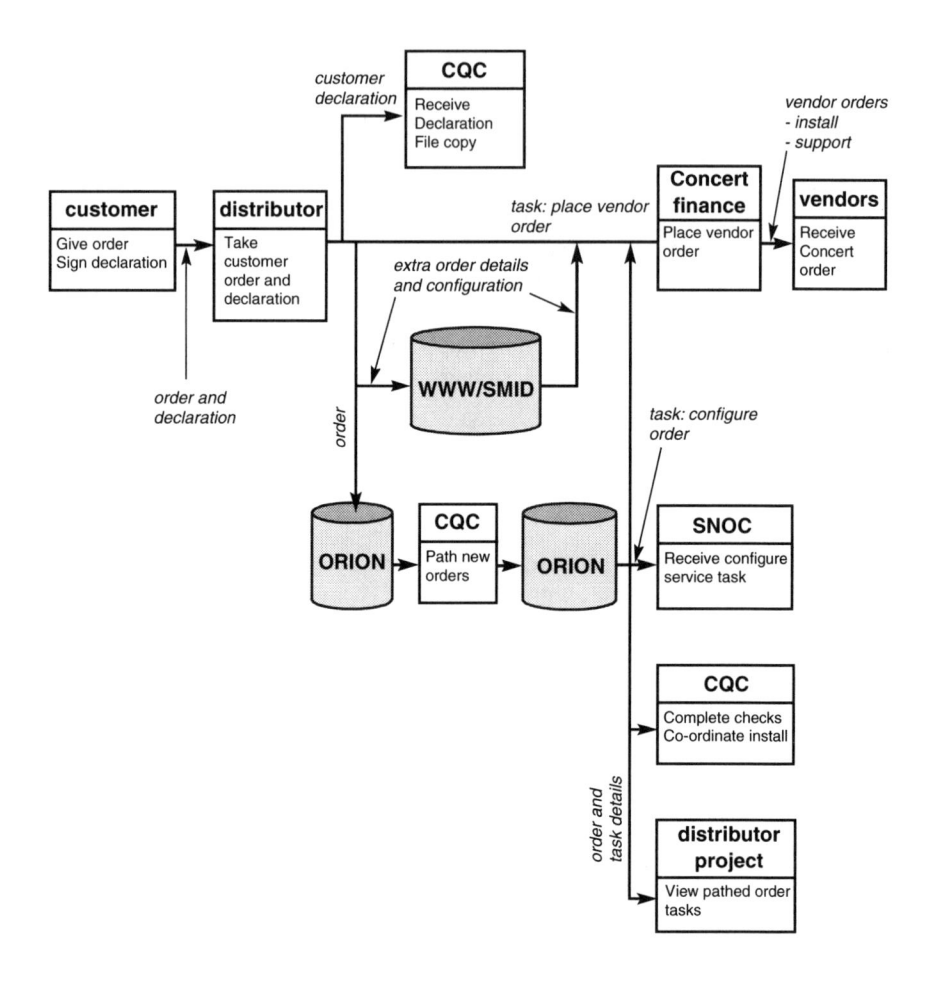

**Fig 12.2** Example process illustration.

## 12.3 ISP Organisational Development and Operations Evolution

Having described what operational models, processes and procedures are, this section now looks at how IP organisations reflect different operational models.

The development of ISPs and the way they operate can be mapped on to the traditional view of the evolution of manufacturing techniques.

### 12.3.1 The Cottage Industry Model

The first 'start-up' ISPs, such as BTnet (BT's initial corporate Internet offering), were like a cottage industry. They utilised highly skilled people, and simple and adaptable systems, while the products were highly tailored to the customer needs. As the operations were centralised and the volumes were low, this was a sustainable model to follow that did not really require the process to be defined. Systems were purely a mechanism for passing and storing data and therefore limited measurements and metrics were produced.

### 12.3.2 The Standardised Products Model

When the product emphasis is still on access products for the corporate IP market, i.e. the ISP is not targeting the mass market, the ISP organisation can move to the next phase on the manufacturing model with standardised products. This means a more scaling-oriented, higher-volume (but still centralised) model can be adopted. Because the products are more standardised, an operational model and processes can be defined. This gives rise to systems development and the definition of measurements and controls. This is also enhanced by the ability to auto-configure some of the more standard services, allowing the more highly skilled workers to concentrate on new product developments.

### 12.3.3 The Lean Production Challenge

In the traditional manufacturing model, it is clear that the next phase is to move to an operations model utilising some of the lean production theories enabling a more mass-production cycle. In the IP world this has not really been the case. Customers still demand a high level of interactivity between the technical specialist in the ISP and their organisations' technical departments. This has been exacerbated by the demands for more comprehensive IP products such as security/firewall, virtual private networks, and global dial and remote access solutions. Combining this with an increase in volume in terms of demand, operations units have expanded rapidly to accommodate this growth. Some of the mass-market products of an ISP, e.g. 'free' dial-up Internet access, have successfully met the lean production challenge, yet for the Corporate IP market the demand from customers for specialisation and expertise prohibits the application of lean production.

### 12.3.4 The Scalable Operations Challenge

This gives rise to a precarious time in the world of operational models and system developments. As IP has become a focus for business drivers and profitability, the strategy demands a move to more scalable operations, looking at existing economies of scale, and a more distributed and less centralised model. With the larger global telcos, this may mean utilisation of existing systems that have delivered economies of scale historically for other products — this could be the right way for the mature and standard IP products, but may not be the answer for the complex IP solutions. As stated earlier, products are now more focused on the value-added and advanced solutions. The data being captured at the front end of the process is more complex, demanding that sales forces have IP skills and that the systems can capture all the relevant data. The technical specification of this product type demands a return to the higher skill sets which are found in 'pockets' within the organisation, and consequently a return to the cottage model for some of the more complex products. Putting this into the context of the global IP network service provider means there is a contradiction in terms of what the company needs to achieve and what the customer wants. Customers want advanced solutions delivered to them in short time-scales, they want it bespoke, they want quality and they want access to the highest skill set in the provider on a 'follow the sun' basis. Contradicting this, the provider wants to standardise products, utilise the existing economies of scale in terms of systems and networks, simplify and automate the complex tasks to reduce resource expenditure, and increase delivery throughput to maximise revenue.

The solution is not clear, but what is certain is that for a global IP network service provider a combination of the three stages of the industrial model need to co-exist within the same organisation. The cottage industry model applies to the centralised, highly skilled teams that are the focus for new and technically complex developments and recently launched solutions. The standardised products model applies to established products that require a moderate degree of customisation. The mass-production model can be applied to the standardised products with some automated procedures that can then move into the leaner, more measured production model. This is an ongoing cycle with products moving from the specialised, highly skilled arena on to the next stages where the technical capability need not be so high. Figure 12.3 shows the flow of people and products over time[1].

### 12.4 People Development

While the organisation continually moves the products along the operations evolution path, the people within the IP environment should develop in the opposite direction, i.e. as the operation of the products are deskilled, the people become more

---

[1] Acknowledgement to Nick Ashley for the skills triangle.

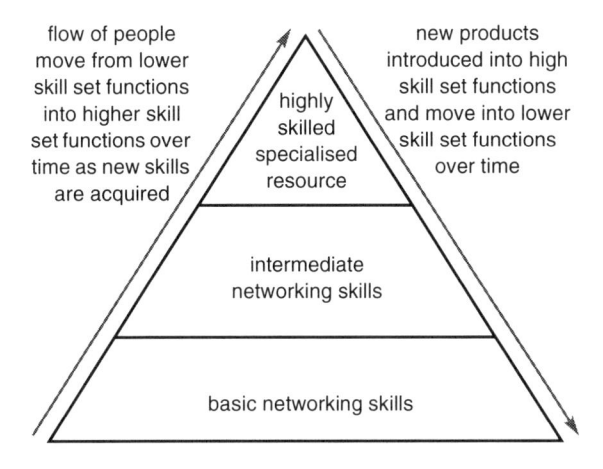

```
                    /\
                   /  \
        flow of people  /   /\ \     new products
        move from lower /   /  \ \   introduced into high
        skill set functions/ highly\  skill set functions
        into higher skill / skilled \ and move into lower
        set functions over/specialised\ skill set functions
        time as new skills/ resource   \  over time
        are acquired    /───────────────\
                       /  intermediate   \
                      /  networking skills \
                     /─────────────────────\
                    /   basic networking    \
                   /         skills          \
                  /_____\
```

**Fig 12.3** Product and people development.

skilled by moving through the organisation into the more development-oriented areas that provide specialist skills.

The people will move on from developing basic networking skills, delivering and maintaining standard products in a controlled and measured environment, into the specialist areas over a period of time. This is a natural and a visible career path to those who want it. It also allows BT and Concert to invest in training without losing its people to competitors.

## 12.5 Conclusions

BT's and Concert's operational models must allow innovation and development, using generic processes where possible and more flexible ones where necessary, but apply strict procedures and measurements where appropriate. Lastly, the systems supporting all of this have to reflect the processes and operational models rather than allowing large and inflexible systems to define the processes. The development lead times for the systems need to be short and flexible to meet the short product introduction cycles and continuous changes in platform design found in BT's and Concert's IP services. BT and Concert strive to meet these criteria in order to be more able to balance the need to be profitable, meet customer requirements and retain scarce skill sets.

# 13

# OPERATIONAL SUPPORT SYSTEMS FOR CARRIER-SCALE IP NETWORKS

## C B Hatch, P C Utton and R E A Leaton

## 13.1 Introduction

Designing and building a large IP network is only a part of creating a carrier-scale IP network. The other significant part, without which no network could be called carrier-scale, is the operational support systems (OSS) — the collection of systems that are used to deliver and manage the network and its services. The effectiveness and the integration of the OSS has a direct impact on the quality of the services and the profitability of the network. In a carrier-scale environment, the OSS has to be a scalable system. With IP networks, new services are continuously being developed with the consequence that the OSS needs continuous redevelopment. This chapter describes the OSS, its relationship to the network and services, and methods for streamlining its design and future development.

The Internet protocol (IP) domain is one of rapid change, vigorous growth and stiff competition. Change is having an impact upon technology, market opportunities and service offerings, vigorous growth is occurring in traffic and numbers of customers, and competition is rife between numerous different Internet service providers (ISPs). In mid-1999 there were around 420 ISPs in the UK [1]. This number continues to change with new entrants to the market and the consolidation of others into larger groupings. BT's goal is to be at the forefront of providing innovative customer services supported by highly efficient operational processes.

BT's growing range of IP services includes fixed and dial access, point-to-point and multipoint connectivity and optional access to the global Internet. The complexity ranges from the simple dial-up service for individual consumers to sophisticated IP virtual private networks (VPNs) for corporate customers with

## Introduction

multiple outstations. Current examples of IP-based services, provided and supported by BT, are shown in Fig 13.1.

The services depicted in Fig 13.1 rely on the delivery of IP 'packets', which is currently on a 'best efforts' basis. There is increasing demand to manage the quality of service (QoS) offered for such services. Indeed, in order to provide 'real-time' services, e.g. voice or video, this will be essential. However, technology to solve this problem is still under development by IP product vendors and many related management capabilities remain to be defined.

Therefore, meeting customer demand and rising expectations in the IP services arena requires significant investment in infrastructure, both in the development of BT's IP network and in the OSS with which it is managed. Indeed, data traffic in the BT IP core is currently growing at an annual rate of 400%. Further evidence of the rapid growth in demand has been a requirement for the tenfold increase in BT's dial IP modem port provision in the past 18 months. BT has also developed wholesale IP products. Consequently, BT needs to manage high volumes of IP traffic which necessitates an effective IP OSS capability, both for the present and in the future.

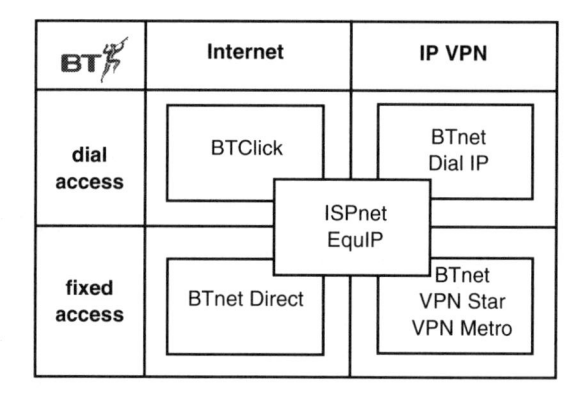

|  BT | Internet | IP VPN |
|---|---|---|
| dial access | BTClick | BTnet Dial IP |
|  | ISPnet EquIP |  |
| fixed access | BTnet Direct | BTnet VPN Star VPN Metro |

**Fig 13.1** Current BT IP-based service portfolio.

The demands of a rapidly changing market for IP services cannot be readily satisfied by traditional telco planning, engineering development and cautious roll-out of services. The design of the IP OSS has to be undertaken within a set of competing pressures, a depiction of which is given in Fig 13.2.

Factors captured in Fig 13.2 include:

- customers' requirements which vary widely with their own knowledge, skills and experience in the IP services and applications areas;

- the drivers on the customer products and services which are subject to frequent changes;

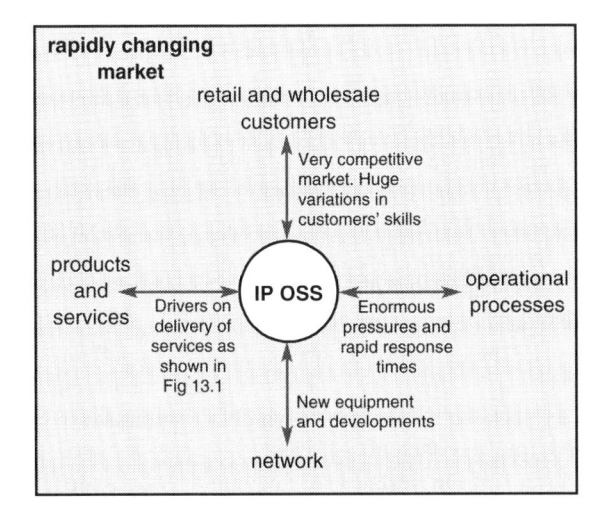

**Fig 13.2**  Rapidly changing market — demands on the IP OSS.

- the operational processes required to provide, maintain and modify services and products;
- the network equipment and related developments, which have to be interfaced with existing systems.

These competing pressures are currently being handled by an IP OSS which is founded on a set of disparate, low functionality systems with a high degree of non-automated management.

Such considerations have led to the need to 'industrialise' the IP OSS to cope with the anticipated demand. This should include optimising the use of automated customer order handling, fault management and associated business-to-business processes, typically using Web-based approaches.

The challenge for BT in the IP OSS design and development area is demanding for reasons that go beyond product, services and technical considerations. These include having to make investment decisions in an environment where the network technology and the vendors supplying it are subject to rapid changes and forecasts of customer demand for access may fluctuate by ± 50% in successive months.

There is short-term competition from ISPs that are not necessarily intent on making a return on capital invested, but are aiming to profit from growth of market share and absorption of both their debts and customers by a larger operator which is investing for the long term. This means that traditional 'business case' planning and due consideration processes for new IP-based services are mostly ineffective or unsuitable in the circumstances now prevailing. Taking this into account, the chapter describes some of the thinking and challenges faced by BT in the quest to design IP OSS for future operations.

## 13.2 The Current Approach to Provision of IP Services

Historically, the BT IP domain has been characterised by a number of small-scale, tactical OSS solutions designed to support specific services — small scale because the level of customer demand was small scale, tactical because they were designed to satisfy the instant demands of an emerging market. They supported specific services as the overall IP and Internet market was evolving rapidly and the properties of a 'generic' service, and consequently its OSS, were not well understood. The disadvantage of this is that they tended to be regarded as 'stove-pipe' solutions that did not interwork particularly well.

As a consequence, the current solution for IP service configuration management has evolved as shown in Fig 13.3. It links essential management systems with bespoke interfaces between them (few of which are automated). Streamlining this complex situation into a coherent, efficient, highly automated and future-proof system is a major part of the design goal and engineering realisation of the future IP OSS.

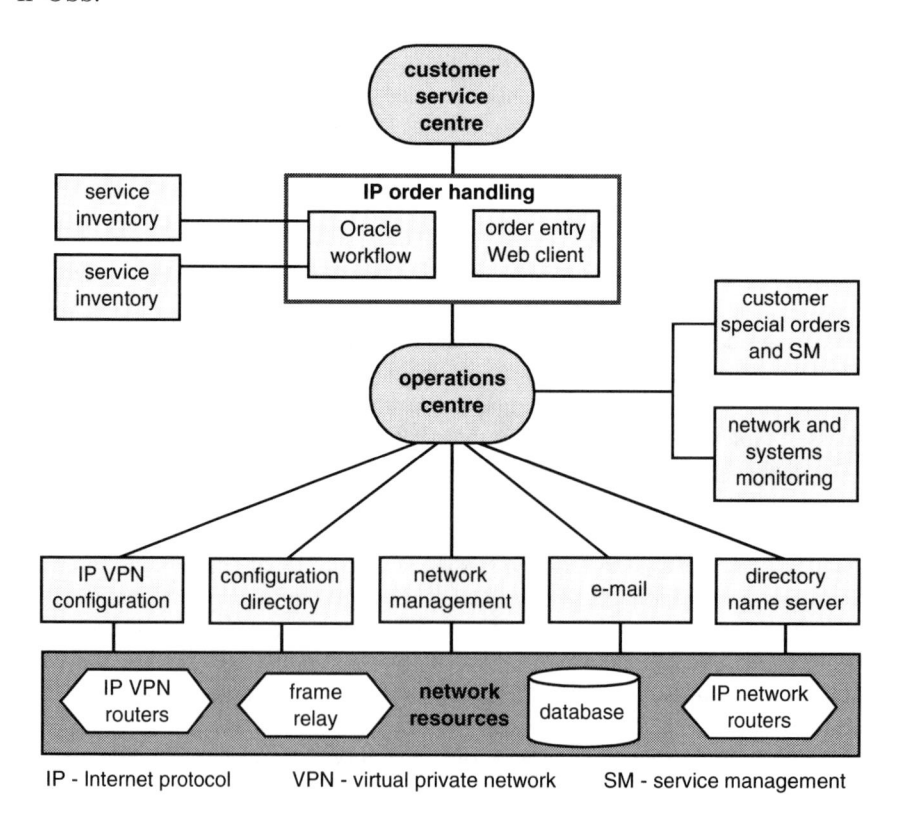

**Fig 13.3** Current view of IP OSS for service configuration.

The operations centre in Fig 13.3 currently co-ordinates various service platform relationships and associated network technologies in order to provide IP services to customers. This situation is illustrated in Fig 13.4, where the multiple IP services shown generally correspond to those introduced in Fig 13.1.

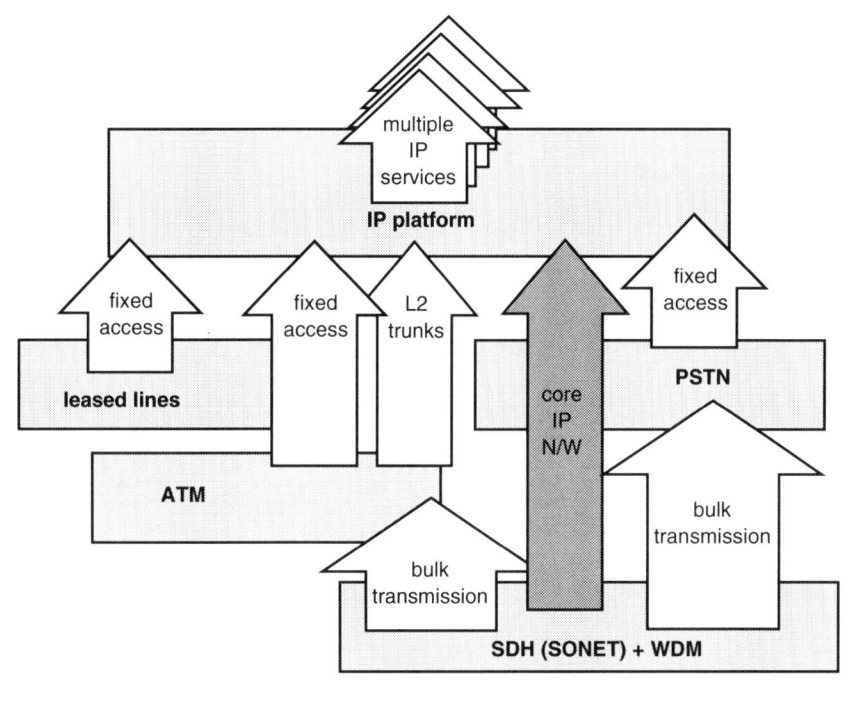

IP - Internet protocol
N/W - network
PSTN - public switched telephone network
ATM - asynchronous transfer mode
L2 - layer 2
SDH - synchronous digital hierarchy
WDM - wavelength division multiplexing
SONET - synchronous optical network

**Fig 13.4** IP OSS service platform relationships and network support.

Figure 13.4 depicts the 'core IP network' in relation to other BT network platforms and technologies. The core IP network is being developed to provide considerable capacity in the UK, linked with proportionate amounts of capacity to provide for global connectivity demands [2].

The key elements discussed so far are networks, systems, people and processes. The other key constituent of OSS is data. The nature of OSS data is complex and highly interconnected. The complexity of OSS data is illustrated in the example shown in Fig 13.5.

Figure 13.5 shows key data entities involved in the many management functions needed to configure and support IP services. It also illustrates why tight data

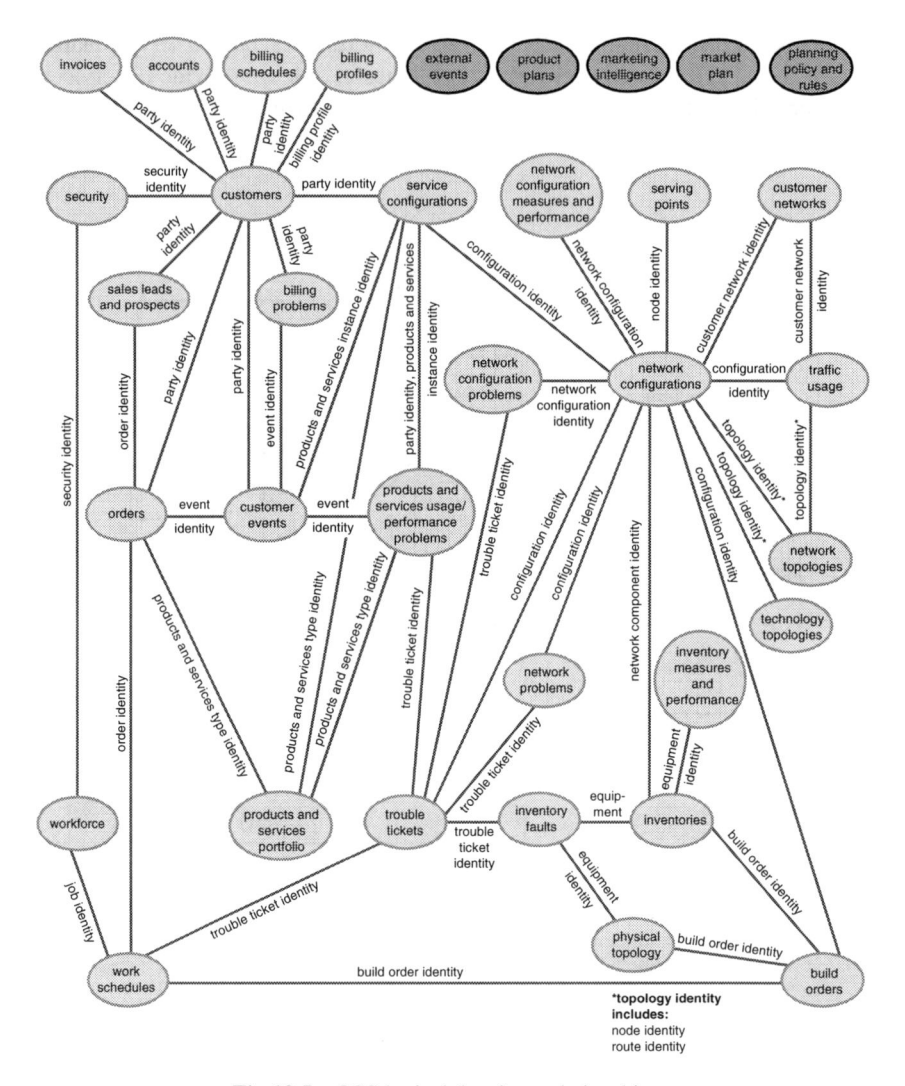

202  *The Current Approach to Provision of IP Services*

**Fig 13.5**  OSS logical data interrelationships.

definition is required as part of the definition of any future OSS architecture. Consolidation of the various platform, data and network management requirements, illustrated in Figs 13.3, 13.4 and 13.5, into an effective solution for management of IP services for the foreseeable future is the immediate design and development challenge for the IP OSS. Furthermore, the IP OSS and the associated processes, as illustrated in Fig 13.5, must be designed to protect the important data and its associations. The strategic direction, as well as some of the technical issues related to this objective, are considered in the next section.

## 13.3 Trends in IP OSS Design, Development and Implementation

The major strategic requirement for the IP OSS is to cope with the anticipated service volumes, primarily for orders but also for requests for service changes. Other key requirements identified at the network management level include the capability to:

- handle significantly increased volumes of network configurations, events, alarms, statistics, etc;
- ensure rapid introduction of new products and services or changes to existing products and services;
- support cross-technology operations;
- support a wide range of IP network elements and network element managers — not just those that may comprise the IP network of today;
- support the management of multiple traffic types/classes supporting different applications in a common network;
- provide support for advanced features such as policy-based management and directory-enabled networks;
- improve the quality of service to the customer;
- reduce operational costs through increased automation;
- migrate product-specific operations to mainstream service operations;
- support interfaces for third-party service providers;
- support migration from the existing OSS and processes.

From a customer service and business management perspective, the following capabilities are important:

- customer on-line access for self-provision of products and services;
- end-to-end automation for provisioning and configuration of IP network services;
- proactive fault management, i.e. to be aware of faults before customers report them;
- improved customer reports;
- integration with other eServices.

Figure 13.6 reflects the capabilities identified in the bullet lists above. It is a 'concept diagram' which is trend-oriented rather than time-specific, but it indicates various factors which are likely to have an impact on the IP domain in the next 2 or 3 year period.

To face up to the anticipated challenges of the next few years, the main driver on OSS development is to redesign the network and service management capability to be optimally fit for purpose. This includes requirements to provide secure open

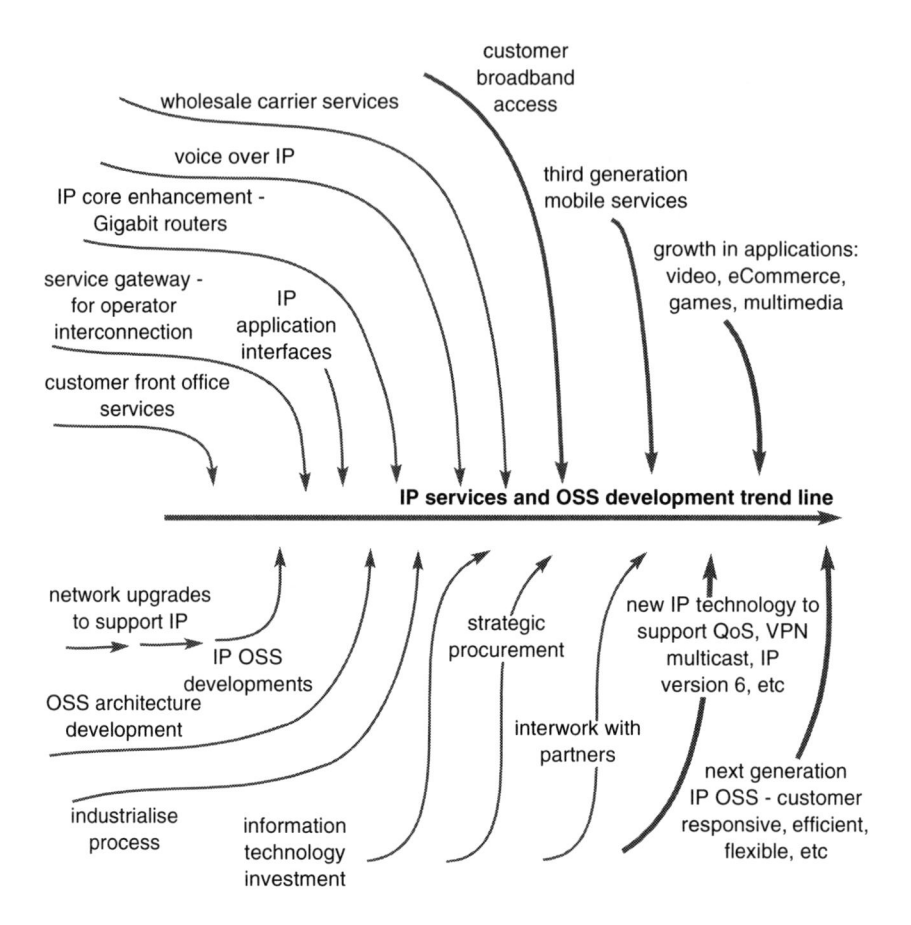

204  *Trends in IP OSS Design, Development and Implementation*

**Fig 13.6**  Factors influencing IP services and IP OSS.

interfaces to all support system components and to introduce a coherent and consistent OSS framework. The OSS domain is currently undergoing a number of paradigm shifts in design outlook to transform it into a fleet-footed engine of change for the future. These shifts are embodied in the following maxims:
- buy not build;
- configure not construct;
- rationalise not proliferate.

BT is not unique in facing these challenges. Other telcos share a similar heritage and are also in search of a next generation OSS which will offer integrated solutions. Observers have identified three generations of OSS in typical industry practice — the first generation being an in-house, monolithic development, the second generation involving numerous disparate systems as a consequence of adopting 'buy-not-build' policies in the absence of an architectural framework (this is where most operators are today), and the third or next generation offering an integrated solution. Clearly, some suppliers would be delighted to offer operators an 'OSS in a box' solution, but, in the IP domain, as no single supplier has a mature, full 'end-to-end' solution, the best prospect for an integrated solution lies in assembling components that will work effectively together. The term 'plug-and-play' is increasingly being adopted to characterise the objectives of next-generation component-based OSS architectures. Current initiatives include that of the Telemanagement Forum (TMF) [3], which is striving to define the 'building blocks' for future OSS.

## 13.4 Design Considerations for an IP OSS

The design of the IP OSS should provide BT with the means to efficiently and effectively manage its IP-based services by aggregating the control of all essential processes needed for the provision of such services. This control should include automatic assignment and configuration of network components in order to provide resources and information to the customer in all key aspects.

One possible approach to the design is based on an emerging model, developed by the TMF [4], in the area of process and function partitioning. This model assembles the individual process flows for service provision which, in turn, may be aligned with the well-established Telecommunications Management Network (TMN) layered management hierarchy defined in M.3010 [5] (see Fig 13.7). The TMN layers may be summarised as follows:

- the business management layer has legal, technical and regulatory responsibility for the total enterprise;
- the service management layer is concerned with, and responsible for, the contractual aspects of services that are being provided to customers or available to potential new customers — some of the main functions of this layer are service order handling, complaint handling and invoicing;
- the network management layer has the responsibility for the management of a network as supported by the element management layer;
- the network element management layer manages each network element on an individual or group basis and supports an abstraction of the functions provided by the network element layer.

Examples from the TMF have the potential to develop into industry-standard reference models against which procurement can be made. If the market in OSS components is to develop and mature, operators and suppliers need to agree on models and frameworks that can be used for this purpose.

The process flows in Fig 13.7 are necessary to ensure the FAB (fulfilment, assurance and billing) model:

206  *Design Considerations for an IP OSS*

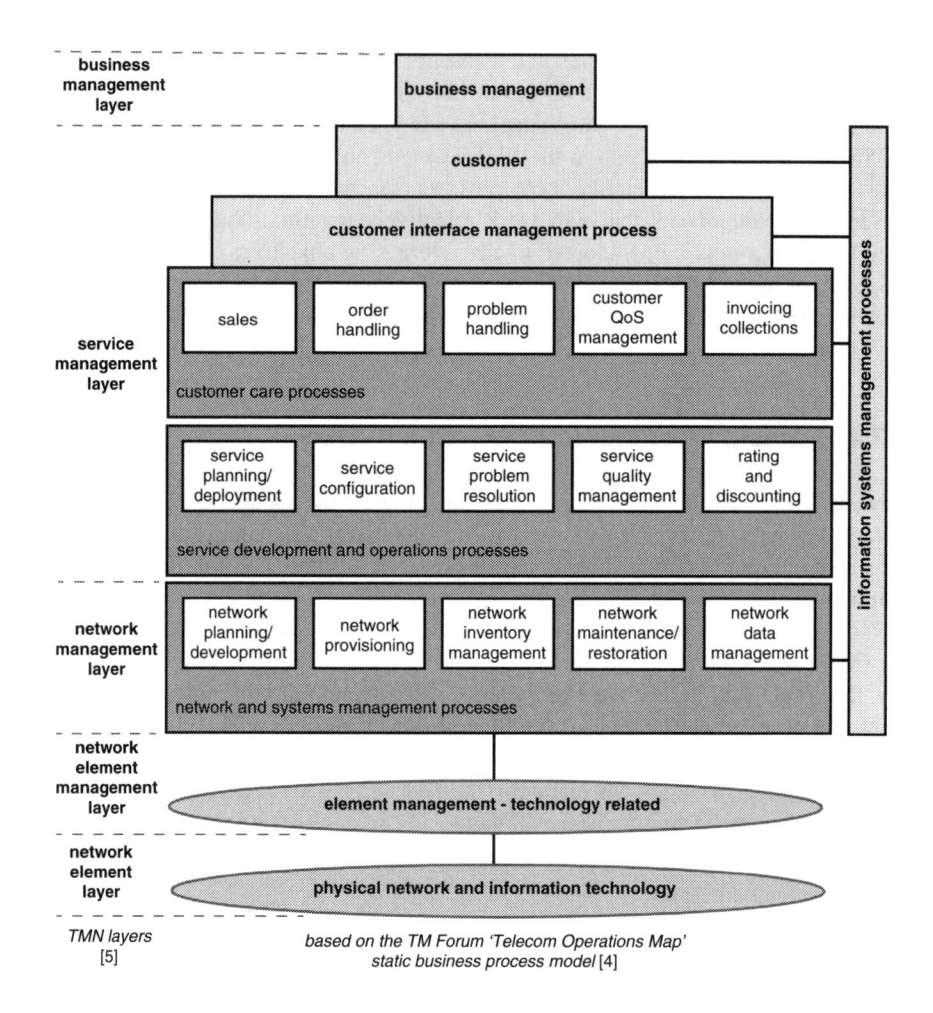

**Fig 13.7** Organisation of process flows in relation to the TMN layer model for the design of an IP OSS.

- service fulfilment — delivery of the service ordered by the customer;
- service assurance — maintaining all aspects of the service provision to the agreed quality;
- service accounting, charging and billing — arrangement of timely, fair and accurate transfers of revenues for the service provided.

A key design consideration for the 'next generation' OSS involves realising a vision for data (i.e. information) as a corporate business resource in itself, as distinct from the systems which may store it at any point in time. The vision includes:

- separation of data, function and process — typically the longevity of the definitions for an element within each of these domains varies from the relatively long lived (i.e. data), to that which changes regularly, typically within each system release (i.e. function), to the particularly volatile (i.e. process) — maintaining a separation between these concerns will preserve investment in the corporate data resource, facilitate reuse of component parts and enable more rapid service deployments;
- logically a single data repository — not necessarily a single physical database but certainly a single logical data model to ensure managed replication and avoid fragments of disconnected, duplicated data;
- uniform and consistent presentation format — independent of native format/location;
- accessible from anywhere, at any time, with the appropriate authority;
- safely and easily managed.

A further category of design consideration, particularly relevant to the topic of carrier-scale IP networks, is that of scalability. Larger scale networks bring increasing demands from a number of perspectives, for example:

- the amount of configuration needed on each network device increases with network size, i.e. the configuration required per node increases with the number of nodes and the number of features enabled (e.g. QoS);
- both the amount of data to collect and the degree of interaction with devices increase with network size, which means increased management traffic and increased processing demands — to handle this effectively requires more distributed processing of management data, and increasingly there is a need for 'smarter' devices which are able to analyse and aggregate data at a local level.

Management of 'smart devices' in an IP network may be achieved using the enhanced access control and security offered by the simple network management protocol version 3 (SNMPv3) [6, 7]. SNMPv3 is a key development in Internet management technologies. Given the new features available, management of large-scale IP networks via SNMPv3 can now become a reality. Network monitoring can be performed in a secure way, protecting the integrity and confidentiality of the information. More importantly, these enhancements bring the option of secure remote hardware configuration. (Security is a recognised shortcoming of both the SNMPv1 and SNMPv2c frameworks.) In addition, SNMPv3 provides enhanced mechanisms and support for building distributed management systems.

The work in the SNMPv3 family of standards is nearing completion, with only minor details being modified, mainly as a result of vendor interoperability and deployment experiments. The gradual adoption of SNMPv3 by the industry should happen during year 2000 with more commercial products coming into the market.

In very large networks, where a large number of devices and ports need to be managed, management traffic may reach unacceptable levels and hence contribute to network congestion. Where out-of-band network management systems are not suitable, BT has developed techniques which can effectively reduce the amount of polling required in the network by varying the polling intervals depending on the degree of change of the values being monitored. Utilising wavelet theory, it works on the simple premise that stable situations require less frequent polling [8].

Building on some of these considerations, the next section describes a vision of an advanced IP OSS design in the context of a proper (i.e. scalable) distributed computing infrastructure within a coherent OSS framework and designed with open OSS interfaces.

## 13.5 Evolution of IP Networks and Services Architecture

As indicated in the preceding sections, the future IP OSS should be:

- process driven;
- component based;
- data (information) centric;
- e-enabled.

To do this, it should exploit maturing technologies in the areas of workflow, middleware and enterprise application integration (EAI).

Workflow technology is the prime means to achieve a process-driven OSS. Workflow concerns the tasks, computer systems and organisational units or people involved for each step in a business process, and the information required as input and to be produced as output. Various companies make workflow automation products that allow a workflow or process model to be defined, along with components such as on-line forms, which can then be used to drive a workflow engine to enact the workflow as a means to manage and enforce the consistent handling of work. Workflow-enabled systems are therefore able to flexibly support a rapidly changing set of business processes. A reference model for this technology, devised by the workflow management coalition [9], is shown in Fig 13.8. Workflow can be applied in a layered fashion to co-ordinate activities from the enterprise level down to individual applications which may contain their own embedded workflow engines.

Middleware provides an environment for distributed applications. From the programmer's viewpoint, it creates the impression of a single virtual machine. Enterprise middleware aims to provide a consistent environment across an entire enterprise. BT, along with many other operators, has a heterogeneous computing environment for both historic and continuing good reasons (i.e. no single package or box does everything).

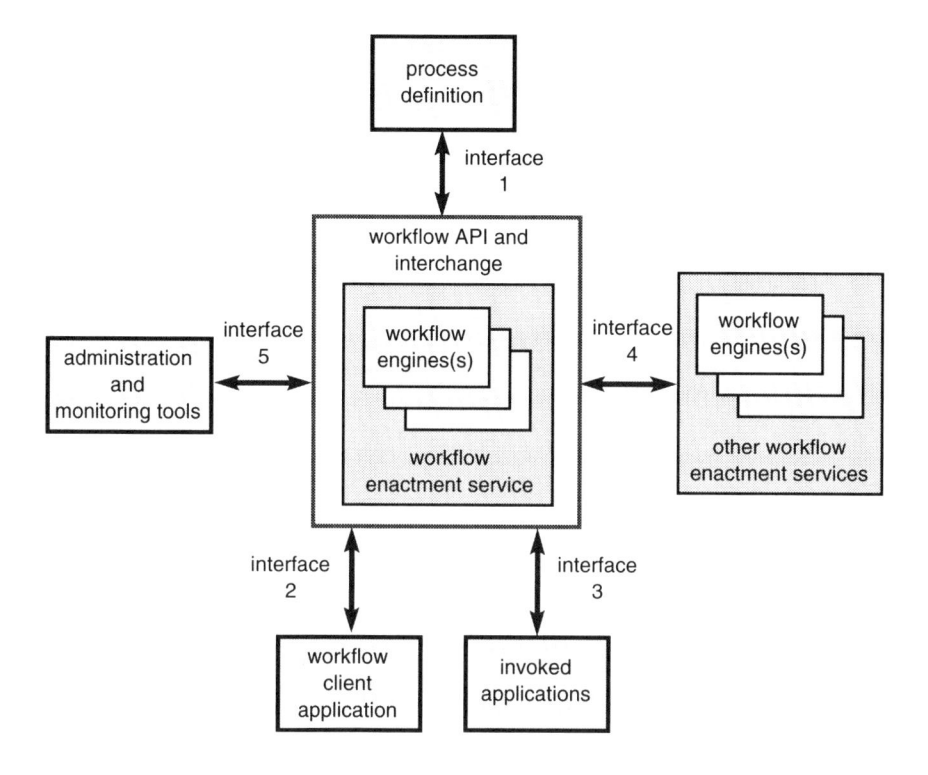

**Fig 13.8** Workflow management coalition reference model.

Unfortunately, as different vendors' middleware products frequently do not interwork, a pragmatic solution would be to use a single vendor's middleware as the 'glue' in order to connect together heterogeneous platforms.

In general, middleware products are designed to interwork with heterogeneous platforms even if (perversely) they do not interwork with other middleware products. For example, CORBA is a standard but its use in application development does not guarantee interoperability.

Enterprise-wide middleware should be regarded as an infrastructure investment. The trend is towards multi-purpose middleware (see Fig 13.9). This provides a variety of interfaces over a layer of common services.

Application integration is now emerging as a topic in its own right, with the appearance of commercial products labelled as EAI or 'integrationware'. Industry groups such as the Open Applications Group (OAG) are developing standards to help achieve 'plug-and-play' business software integration.

Various publications appearing from industry consultants [10–13] are also raising its profile.

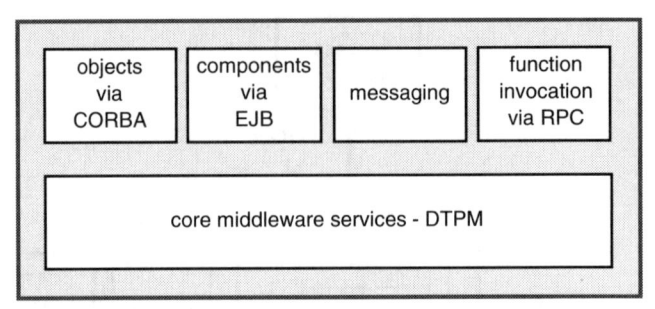

210  *Evolution of IP Networks and Services Architecture*

```
┌─────────────────────────────────────────────────────────────┐
│  ┌──────────┐ ┌──────────┐ ┌──────────┐ ┌──────────────┐   │
│  │ objects  │ │components│ │          │ │  function    │   │
│  │   via    │ │   via    │ │messaging │ │  invocation  │   │
│  │  CORBA   │ │   EJB    │ │          │ │   via RPC    │   │
│  └──────────┘ └──────────┘ └──────────┘ └──────────────┘   │
│  ┌─────────────────────────────────────────────────────┐   │
│  │        core middleware services - DTPM              │   │
│  └─────────────────────────────────────────────────────┘   │
└─────────────────────────────────────────────────────────────┘
```

CORBA - Common Object Request Broker Architecture
DTPM - distributed transaction processing monitor
EJB - Enterprise JavaBeans
RPC - remote procedure call

**Fig 13.9**  Multipurpose middleware.

The Butler Group [10] defines application integration as:

'The requirement to integrate into new business processes the functional behaviour or business rules of disparate systems as well as, but not just, the data that underlies them.'

A 'buy-not-build' policy will necessitate an increasing focus on applications integration where the applications are likely to include a mixture of third-party packages from different sources, legacy systems and some continuing new build in specific areas which offer distinct competitive advantage.

A particular goal of modern application integration is real-time integration, i.e. the ability to bring together and process information from disparate sources in real time. This offers significant potential benefits for supply-chain optimisation. In the context of Web-enabled customer interfaces for IP services, this means being able to check the disposition of the network before a customer's order is accepted and a commitment is made to deliver a service which proves to be incompatible with their local network or equipment.

Today's integration practice is broadly on a level with glue and sticky tape. A goal of integration is to be non-intrusive and require minimal effort (especially programming). Off-the-shelf translators and adapters can be useful tools although they can still need lots of manual configuration, e.g. to define parameter mapping.

In the longer term, the OAG has proposed the concept of an 'integration backbone' as a 'bus' style of interconnection (of complexity order $n$) to contrast with a 'mesh' style of interconnection (of complexity order $n^2$) which tends to pervade second generation OSS, as shown in Fig 13.10. The OAG has further 'bus' style interconnect    'mesh' style interconnect

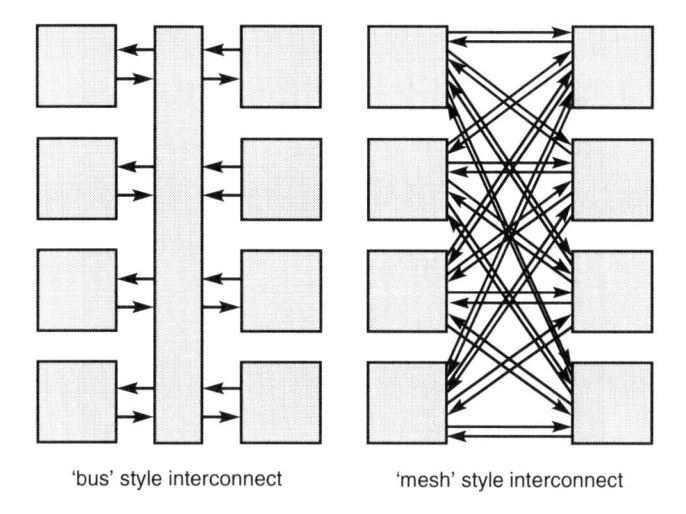

**Fig 13.10**  The integration backbone concept.

proposed OAGIS (OAG Integration Specification) as a standard for plug-and-play integration [14].

Integration is the bringing together of incompatible resources, but it is the lack of common business and technical architectures that makes integration complex. It is not just the difference in data formats and interfaces, but the lack of a common definition of business concepts that causes problems. Common architectures and standards are a key requirement.

The Butler Group advocates a three-layer approach to defining architecture, which is depicted in Fig 13.11. This group recommends that larger enterprises establish architectures and infrastructures that enable rapid integration to become 'business as usual'.

Building on all the ideas outlined so far, the latest vision for BT's OSS architecture is depicted in Fig 13.12.

A description of the components and functions of the architecture shown in Fig 13.12 follows.

- System architecture and workflow functions:

    — 'enterprise workflow' (EWF) controls transactions that affect the whole business, not individual components of that business;

    — the enterprise middleware/enterprise application integration (EMW/EAI) component of the architecture provides the underlying information technology infrastructure which handles the distribution, security, transactional integrity, etc, between the functions;

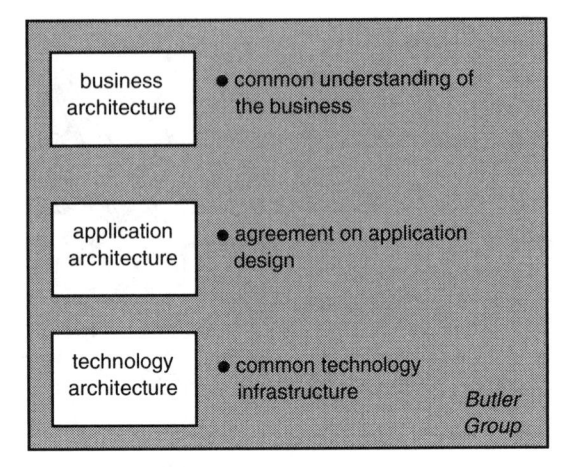

**Fig 13.11** Architecture layers.

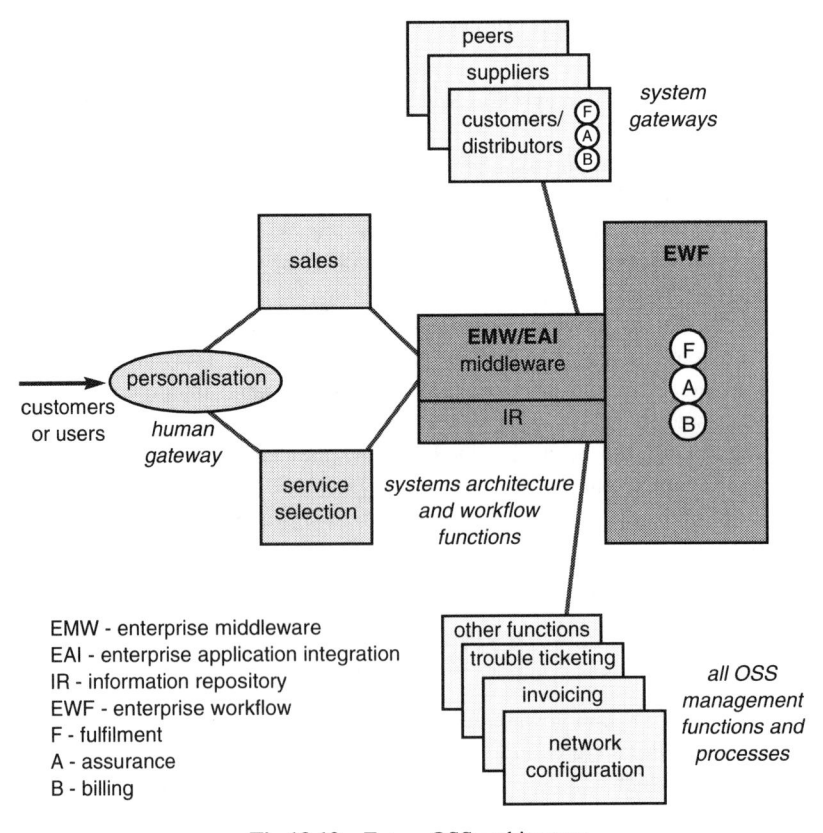

**Fig 13.12** Future OSS architecture.

— the EWF controls all the end-to-end transactions in the enterprise, but the key customer-related ones can be classified simply as fulfilment, assurance and billing (this corresponds to the FAB (fulfilment, assurance and billing) process flows described in section 13.4, relating to Fig 13.7).

- Information repository:

    — the information repository (IR) provides access to data, including location and transformation services, and implements the agreed information architecture (IA);

    — the IA provides a common information model which links together all the technical and commercial informational elements in a single viewpoint;

    — it provides a high-level, logical viewpoint of the information constructs pertaining to the operation of the whole business;

    — the IR and its implementation of the IA are fundamental to the operation of the architecture; successful IP businesses recognise that information is a key strategic asset because it is the currency of electronic business.

- System gateways:

    — a fundamental part of the architecture is a set of system gateways to interface to other businesses, because where there is an interaction required with another legal entity, the only acceptable route is through a gateway;

    — the gateways will support the key fulfilment, assurance and billing processes;

    — higher level interfaces will be based increasingly on extensible markup language (XML) and eCommerce libraries, while lower level interfaces, for example those used for network element management such as simple network management protocol (SNMP), will be supported where appropriate, i.e. within IP OSS management functions.

- Human gateways:

    — as well as system gateways, there will be a set of gateways to allow people to interact with the system, rather than just machine-to-machine interactions;

    — the 'human gateway' is currently seen as comprising three components — a system to ensure that users see a personalised interface and systems to allow sales and service selection.

- IP OSS management functions:

    — in order to provide the required IP OSS functions and processes, there should be a number of suitable components provided to support these requirements, linked by the central 'backbone' capabilities of EWF, EMW/EAI and IR;

— this set of functions and processes could be organised in accordance with an emerging architecture, such as that presented in Fig 13.7.

To conclude, an example of use of the system architecture, presented in Fig 13.12, could be as follows.

A customer places an order by telephone for an all-inclusive IP service. The following processes should occur:

- a service agent activates a fulfilment workflow;
- the EMW/EAI establishes a sequence of activities;
- IP OSS functions are called to fulfil the service request;
- the IR collects, distributes and manages the necessary information transactions;
- electronic gateways handle orders to peer suppliers for access connections, PC delivery or any other service aspect needed to complete the customer's requirements (peer suppliers may include, for example, a service and applications provider, core IP provider, mobile access provider, core transport provider and/or fixed access provider, all interconnected such that all these separate business entities interact through their service gateways).

The current status of the architecture is that the overall shape has been widely agreed and there is now considerable activity planned in the areas of interface definitions (both the system and human) as well as proving the underlying technology in the IP OSS domain.

## 13.6 Conclusions

This chapter has discussed various issues which are influencing BT's views in the area of IP OSS design and implementation. Such an OSS will become a major part of BT's infrastructure, providing vital resources for future business development, which is expected to be strongly dependent on IP-based networks and services. The discussion has been set in the context of associating the current and prospective BT IP products and solutions with the underlying demands for new services, improved QoS and highly efficient methods for provision and management of such services.

Moreover, the IP OSS has been described in terms of strategic drivers and design trends and also within a broader vision of architectural evolution appropriate to future product, service and customer solutions.

## References

1 The Durlacher Quarterly Internet Report (Q2 1999) — http://www.durlacher.com or http://www.intellact.nat.bt.com/intellact/reports /durlacher/main.htm

2   BTnet Services: '*Our network*' homepage and linked information — http://www.bt.net/network.htm

3   TeleManagement Forum — http://www.tmforum.org/

4   TeleManagement Forum, GB910 '*Telecom operations map*', evaluation version 1.1 (April 1999) — http://www.tmforum.org/documents/gb910.pdf

5   ITU-T Recommendation M.3010: '*Principles for a telecommunications management network*', (May 1996).

6   IETF RFC 2570, '*Introduction to version 3 of the Internet-standard network management framework*', ftp://ftp.isi.edu/in-notes/rfc2570.txt (April 1999).

7   IETF RFC 2571: '*An architecture for describing SNMP management frameworks*', http://www.ietf.org/rfc/rfc2571.txt?number=2571 (May 1999).

8   Teo S A, Davison R, Domingos J and Malenchino V: '*A self-adjusting frequency polling technique for IP networks*', submitted for the IEEE workshop on IP-oriented Operations and Management 2000.

9   Workflow Management Coalition Reference Model — http://www.aiim.org/wfmc/standards/model2.htm

10  Butler Group: '*Application Integration Management Guide — strategies and technologies*', (June 1999) — http://www. butlergroup.com/

11  Highsmith, J.: '*Application integration — high risk or high reward?*', (August 1999) — http://www.cutter.com/consortium/

12  Datamonitor: '*Enterprise application integration (EAI) — connection or integration?*', (July 1999) — http://www.intellact.nat.bt.com/intellact/reports/datamonitor/index.htm .

13  Linthicum, D. S.: '*Enterprise Application Integration*', Addison Wesley (November 1999).

14  OAG White Paper: '*Plug and play business software integration*' — http://www.openapplications.org/

# 14

# IP ADDRESS MANAGEMENT

## P A Roberts and S Challinor

## 14.1  Introduction

Networks need addresses to identify objects and locations in the network. The structure of the address is the single largest influence on the nature of the network. To fully understand a network you must understand its address structure. This chapter describes in technical depth the structure of Internet protocol (IP) addresses and how these addresses are managed.

IP addresses are the unique numbers used to identify individual connections to the Internet; they are also a limited resource and the careful management of these IP addresses is essential to the running of the Internet. BT must be able to efficiently manage its own and its customers' IP addresses and it must be able to demonstrate good management of IP addresses in order to qualify for more IP address space. Without more new IP addresses BT would not be able to expand its IP network or launch new IP services. It is therefore of central importance to BT that it manages its IP addresses well.

The chapter starts with a description of what IP addresses are, before going on to outline the global IP address environment and how the global address space is managed through regional registries and organisations, such as Internet service providers (ISPs), for distribution to end users. This is followed by an examination of the process for obtaining addresses for an end user which highlights some of the rules which have to be followed by the ISP as part of this process. The penalties for abuse of this system are punitive and could seriously affect the ability of an ISP to conduct its business.

## 14.2  IP Addresses

An IP address consists of 32 bits, written as four 8-bit numbers (thus in the range 0 to 255) separated by a decimal point. Every IP address is invisibly divided into two parts, the first part identifies a particular network, and the second part identifies a particular machine on that network (called a 'host'). The point at which the first part, the network part, of the address ends and the second part, the host part, of the address begins depends upon the size of the network. If the network is very large, with many hosts, the division point is nearer the beginning of the address than it would be in a smaller network with fewer hosts. For instance, with the address 132.146.100.100, the first two numbers (132.146) may refer to a single large network, leaving address space from 0.0 through to 255.255 (i.e. approximately 65 000 addresses). Whereas the address 194.72.10.10 may be the address of a host on a smaller network represented by the numbers 194.72.10 which leaves only 255 addresses on that network (0 to 255).

The exact position of this division point is denoted by the 'subnet mask', which indicates which part of the address represents the network, and which represents the host. The 132.146.100.100 address would have a subnet mask of '255.255.0.0', the 255s indicating which numbers represent network, and the zeros indicating which numbers represent the hosts on that network, the 194.72.10.10 example would have the subnet mask '255.255.255.0'. What is in fact happening here is that every bit in the 32 bits of the address that represents the network is being masked with a '1', and every bit that represents a host is being masked with a zero. Thus the subnet mask for the first example is actually 11111111.11111111.00000000.00000000, but for convenience this is converted to decimal and written as 255.255.0.0.

The first address on a network cannot be given to a host on that network, since it is used to represent the network as a whole. For example, the address 194.72.10.0, with a subnet mask of 255.255.255.0 is the first address on that network, and is used only to represent that entire network. This is called a 'network' address. Similarly the last address on a network cannot be given to the host, since this is used as a broadcast address. Any traffic sent to a broadcast address is effectively addressed to all the hosts on that network. The address 194.72.10.255 with subnet mask 255.255.255.0 is a broadcast address.

The original Internet plan was to break the whole address space into different sections to allow for the creation of these different sized networks; this plan is no longer used, but the terminology from it still is — it is presented below:

Class A: 0-127.X.X.X

126 Class As exist (0 and 127 are reserved)
16 777 214 hosts on each Class A

Class B: 128-191.X.X.X

16 384 Class Bs exist
65 532 host on each Class B

Class C: 192-223.X.X.X

2097 152 Class Cs exist
254 hosts on each Class C

Class D: 224-247.X.X.X

Class Ds are multicast addresses

Class E: 248-255.X.X.X

Reserved for experimental use

The above scheme proved very wasteful of address space, since the smallest network that could be created had 255 addresses. Networks with less than 254 machines would lock up address space, preventing any other network from using it. The scheme of classless interdomain routing (CIDR) or 'variable subnetting' was devised to solve this. With a variable subnet, the network part of the address is extended into the last 8 bits of the address, creating space for more networks, but with less hosts since there are less bits left with which to address hosts. For example, if a network address were represented by the first 26 bits of the 32 bits available, that would leave only 6 bits for the machines on that network.

If the subnet mask for a 26-bit network was written out in binary, it would look like this:

11111111.11111111.11111111.11000000

Converting each of the 8 binary digits to decimal gives the subnet mask for a 26-bit subnet as:

255.255.255.192

Under the original addressing plan, the Class C network 194.72.10.0 would have 255 hosts, but there would only be one network, and any unwanted addresses would be wasted. If this is converted to a 26-bit network, the one class C network is swapped for 4 smaller networks. To see how this is done, only the last 8 bits of the address will be dealt with for the moment. The first 2 bits have been borrowed for use by the network, and with 2 bits only four different combinations can be represented: 00, 01, 10 and 11. Now since these two bits are actually the first two bits in an 8-bit binary number, they actually represent the numbers 128 and 64. Thus the four combinations 00, 01, 10 and 11 are actually written as 0, 64, 128 and 192. The hosts may be written with 6 binary digits, giving 64 combinations, representing the decimal numbers from 0 through to 63. Since the last two bits of the 26-bit network description and the 6 bits of the host's description are to be written as one number, the host's value (0 to 63) must be added on to the decimal value of the network (0, 64, 128 or 192), giving the following four networks to use:

Network 1: 194.72.10.0 through to 194.72.10.63

Network 2: 194.72.10.64 through to 194.72.10.127

Network 3: 194.72.10.128 through to 194.72.10.191

Network 4: 194.72.120.192 through to 194.72.10.255

All have the subnet mask 255.255.255.192.

Since the first address of any network is the network address, and the last is always the broadcast address, there are only 62 addresses available for use by hosts on each of the above networks. A Class C network capable of supporting 254 hosts has been exchanged for four smaller networks, each capable of supporting 62 hosts. It can be seen that the subnet mask is critical in understanding what an IP address is actually representing, and even with the subnet mask it is far from intuitive to determine whether the address is a machine address, a network address or a broadcast address.

The boundary between the network part of the address can be moved further into the last 8 bits, allowing for more and more networks, each with fewer and fewer hosts. For example, a 30-bit network would have only 2 bits available for addressing the machine. Since 2 bits only allows four combinations, and two of those have to be used for network and broadcast addresses, there can only be two machines on this network — which is perfect for addressing a point-to-point link. Thus 30-bit networks are very common within an IP network.

Table 14.1 shows the number of addresses available with a given size network, where the network size is specified in both netmask notation and the '/' (slash) notation which identifies the number of bits belonging to the network part of the address.

## 14.3 Reserved Ranges of IP Addresses

The following ranges have been reserved for use on private networks and are known as private addresses or RFC 1918 addresses, these ranges are not routable across the Internet:

10.0.0.0/8 or 10.0.0.0 — 10.255.255.255

172.16.0.0/12 or 172.16.0.0 — 172.31.255.255

192.168.0.0 or 192.168.0.0 — 192.168.255.255

## 14.4 IP Address Co-ordination

### 14.4.1 Global Co-ordination

The growth of the global Internet from its beginnings as the US Government's experimental ARPAnet soon identified a need to control the allocation of addresses to the many networks being connected. This task fell to an organisation called the Internet Assigned Numbers Authority (IANA) which was set up and funded by the US government. IANA controlled the management of global IP address allocation by delegating the responsibility to regional Internet registries (RIRs), each responsible for a geographic area:

**Table 14.1** Comparison of netmask and / (slash) notation.

| Mask | / | Hosts | Networks | Host bits | Network bits | Addresses |
|---|---|---|---|---|---|---|
| 255.255.255.255 | 32 | 1 | N/A | 0 | 0 | 1 |
| 255.255.255.254 | 31 | 0 | 2147483648 | 1 | 31 | 2 |
| 255.255.255.252 | 30 | 2 | 1073741824 | 2 | 30 | 4 |
| 255.255.255.248 | 29 | 6 | 536870912 | 3 | 29 | 8 |
| 255.255.255.240 | 28 | 14 | 268435456 | 4 | 28 | 16 |
| 225.255.255.224 | 27 | 30 | 134217728 | 5 | 27 | 32 |
| 225.255.255.192 | 26 | 62 | 67108864 | 6 | 26 | 64 |
| 225.255.255.128 | 25 | 126 | 33554432 | 7 | 25 | 128 |
| 225.255.255.0 | 24 | 254 | 16777216 | 8 | 24 | 256 |
| 225.255.254.0 | 23 | 510 | 8388608 | 9 | 23 | 512 |
| 255.255.252.0 | 22 | 1022 | 4194304 | 10 | 22 | 1024 |
| 255.255.248.0 | 21 | 2046 | 2097152 | 11 | 21 | 2048 |
| 255.255.240.0 | 20 | 4094 | 1048576 | 12 | 20 | 4096 |
| 255.255.224.0 | 19 | 8190 | 524288 | 13 | 19 | 8192 |
| 255.255.192.0 | 18 | 16382 | 262144 | 14 | 18 | 16384 |
| 255.255.128.0 | 17 | 32766 | 131072 | 15 | 17 | 32768 |
| 255.255.0.0 | 16 | 65534 | 65536 | 16 | 16 | 65536 |
| 255.254.0.0 | 15 | 131070 | 32768 | 17 | 15 | 131072 |
| 252.252.0.0 | 14 | 262142 | 16384 | 18 | 14 | 262144 |
| 255.248.0.0 | 13 | 524268 | 8192 | 19 | 13 | 524288 |
| 255.240.0.0 | 12 | 1048574 | 4096 | 20 | 12 | 1048576 |
| 255.224.0.0 | 11 | 2097150 | 2048 | 21 | 11 | 2097152 |
| 252.192.0.0 | 10 | 4194302 | 1024 | 22 | 10 | 4194304 |
| 255.128.0.0 | 9 | 8388606 | 512 | 23 | 9 | 8388608 |
| 255.0.0.0 | 8 | 16777214 | 256 | 24 | 8 | 16777216 |
| 254.0.0.0 | 7 | 33554430 | 128 | 25 | 7 | 33554432 |
| 252.0.0.0 | 6 | 67108862 | 64 | 26 | 6 | 67108864 |
| 248.0.0.0 | 5 | 134217726 | 32 | 27 | 5 | 134217728 |
| 240.0.0.0 | 4 | 268435454 | 16 | 28 | 4 | 268435456 |
| 224.0.0.0 | 3 | 536870910 | 8 | 29 | 3 | 536870912 |
| 192.0.0.0 | 2 | 1073741822 | 4 | 30 | 2 | 1073741824 |
| 128.0.0.0 | 1 | 2147483646 | 2 | 31 | 1 | 21474836484 |

- the American Registry for Internet Numbers (ARIN) [1] covers North and South America, the Caribbean and some parts of Africa — ARIN took over this role from Network Solutions in December 1997;
- Réseaux IP Européens (RIPE) [2] covers Europe, North Africa and parts of the Middle East and was formed in 1989 — a number of working groups exist within

RIPE, operating in areas of common interest to the Internet community which are run by volunteers from participating local Internet registries;
- the Asia Pacific Network Information Centre (APNIC) [3] covers the Asia Pacific region.

There have been proposals to set up regional registries for Africa and Latin America which would possibly be known as AFRINIC [4] and LATNIC [5].

IANA was financed by the US government which, due to the global nature of the network, felt the Internet community should take responsibility for financing a top-level management organisation. As a result of this, the Internet Corporation for Assigned Names and Numbers (ICANN) was formed in October 1998. ICANN [6] is a non-profit corporation that not only has responsibility for IP address space allocation, but also looks after protocol parameter assignment, domain name system management, and root server system management functions. ICANN is already operational and it is in the process of assuming the responsibility for the technical management and policy development functions that have been assigned to it. The transition to ICANN from the US Government and IANA is scheduled to be completed by September of 2000. The current structure of ICANN (see Fig 14.1) is composed of a board of nineteen directors consisting of nine 'at-large' directors, nine nominated by ICANN's three supporting organisations, and the President/CEO. The nine current 'at-large' directors are serving initial terms and will be succeeded by at-large directors selected by an at-large membership organisation which will seek to represent the broadest possible spectrum of the Internet community. The three ICANN supporting organisations consist of the:

- Protocol Supporting Organisation (PSO) which is made up of representatives of the Internet Engineering Task Force (IETF), World Wide Web Consortium (W3C), International Telecommunications Union (ITU) and European Telecommunications Standards Institute (ETSI) — the PSO will advise the ICANN board with respect to matters relating to the assignment of parameters for Internet protocols and technical standards;

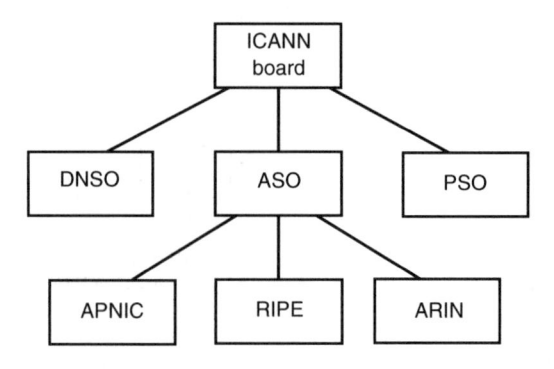

**Fig 14.1** The structure of ICANN with details of the ASO part.

- Address Supporting Organisation (ASO) which consists of representatives from the ARIN, RIPE and APNIC RIRs — the ASO will advise the ICANN board with respect to policy regarding IP addresses;
- Domain Name System Supporting Organisation (DNSO) which consists of representatives from top-level domain registries, commercial and business entities, gTLD registries, ISPs and connectivity providers, non-commercial domain name holders, registrars, and trademark, other intellectual property and anti-counterfeiting interests — the DNSO will advise the ICANN board with respect to policy issues relating to the domain name system.

### 14.4.2 Regional Internet Registries

As outlined in the above section there are currently three RIRs, each of which has responsibility for a different area of the world, each RIR works towards the following goals:

- fair distribution of address space to ensure there is enough address space to go round, and ensuring that end users are considering the use of private address space and that assignments to users are only for the short term needs of their network;
- conservation to prevent the practice of stockpiling of addresses by ensuring that each application for an addresses is genuine and that assignments to users are only for the immediate and short-term needs of their network which are actually going to happen, rather than for long-term projects or growth which may not come to fruition;
- aggregation to create a hierarchical distribution of globally unique address space which will permit the simplification of routing information — this is achieved by the RIRs being allocated very large blocks of address space, typically /8s, and, in turn, allocating large blocks of addresses to the LIRs for onward assignment to their customers, which results in a relatively small number of blocks representing large areas of the world and each LIR will only announce the larger blocks to other ISPs, thus helping to reduce the size of the global routing tables;
- registration, including the provision of public registry or database, to ensure uniqueness and assist troubleshooting — details of all addresses allocated to an LIR are entered into a public database as well as the addresses assigned to end users by an LIR (contact information for these address ranges is also included and this information may be accessed using the 'Whois' tool [7]).

All of the above registries are run on a non-profit basis, charging the participating local Internet registries to cover their administrative costs. Regional registries are also responsible for issuing Autonomous System (AS) numbers and 'in-addr-arpa' reverse DNS delegations.

### 14.4.3 Local Internet Registries

Most LIRs are Internet service providers who require addresses to assign to customers connected to their network. Some organisations that have very large networks or that connect to multiple ISPs can become enterprise registries which, like an ISP, will have an IP address allocation but it is only for use within the organisation. The LIR will announce the block of addresses allocated to it to the Internet using the BGP exterior routing protocol. BT runs an enterprise registry for its internal address space and a public registry for its ISP BTnet and a number of registries in the joint ventures and Concert.

## 14.5 IP Assignment Requests

When an LIR first starts up it is given a limited number of addresses as its first allocation and the first assignments from this block have to be sent to the RIR for approval. As the LIR demonstrates competence in this activity, then it will be given more power to assign address space independently. In order to achieve this, APNIC and RIPE allocate each LIR an assignment window; this is a representation of the number of IP addresses the registry may assign to an end user without first seeking approval from the RIR. The assignment window is usually expressed as a network mask in '/' notation. The use of the terms 'assign' and 'allocate' are important and should not be confused — an RIR allocates addresses to an LIR for onward assignment to end users.

An LIR has to follow the rules laid down by the RIR when they assign IP addresses to an end user. Only an LIR can assign IP addresses to an end user; an LIR cannot assign an IP address to a third party for onward assignment to the end user. Failure to abide by the rules could result in the LIR having its assignment window reduced to zero, meaning that all requests for assignment would have to be processed through the RIR, resulting in delays to the customer's IP address application process.

An end user applying for address space from an LIR will be expected to provide the following information prior to an assignment being made:

- overview of end-user organisation describing the structure of the organisation requesting the addresses and in which part of that organisation they would be used, including information on:

  — how the addresses will be distributed among the various departments;

  — the geographical set-up of the organisation;

  and, if the organisation is a subsidiary of another organisation or it has any subsidiaries, address usage of all parts of the organisation will have to be considered as part of the application;

- contact details for a technical representative which will be entered into the RIR's public database;
- contact details for an administrative representative, who must be someone on the site where the IP addresses are used — this data will also be entered into the RIR's public database;
- an addressing plan showing how the organisation is using all public address space that has been assigned to it in the past and what its future plans are for that space (see Fig 14.2);
- an addressing plan showing how the requested IP addresses will be used over the next two years (see Fig 14.3) — RIPE expects that 80% or more of the IP addresses requested will be used within two years of assignment while ARIN expects 50% or more of addresses requested will be used within one year of assignment.

#[CURRENT ADDRESS SPACE USAGE TEMPLATE]#

| Prefix | Subnet Mask | Size | Addresses Used Current | 1-yr | 2-yr | Description |
|---|---|---|---|---|---|---|
| 194.72.193.0 | 255.255.255.192 | 64 | 28 | 34 | 60 | Department A |
| 194.72.193.64 | 255.255.255.224 | 32 | 10 | 21 | 28 | Department B |
| 194.72.193.96 | 255.255.255.224 | 32 | 8 | 13 | 27 | Department C |
| 194.72.193.128 | 255.255.255.128 | 128 | 0 | 0 | 0 | Unused |
| 194.72.194.0 | 255.255.254 | 512 | 317 | 429 | 480 | Department D |
| | | 640 | 363 | 497 | 595 | Totals |

**Fig 14.2** The 'Current Address Space Usage Template' is used to show how addresses are currently used within the organisation.

The LIR is expected to keep records of this information in case of an audit by the RIR and forward to the RIR if the request is greater than their assignment window. In Europe this would be submitted to RIPE using the RIPE-141 form, an example of which is shown in the Appendix. The form is intended to present enough information to the RIPE hostmaster to make an informed decision about the request.

Once the form is completed it is e-mailed to RIPE via a syntax checking robot and is allocated a ticket number which is mailed back to the requester. The request is then placed in a wait queue before being passed to a member of the RIPE staff known as a hostmaster, who may then enter into an e-mail conversation with the requester if they require any further details. It is in the interest of the requester and the end user to ensure all information presented is correct and clear in order to avoid any unnecessary delays in this stage of the application. The hostmaster has to consider the request in terms of the goals of the RIR/LIR system. Once the request is authorised the LIR will assign the address space from their existing block. Before the end user starts using the addresses the assignment should be entered into the RIR's public database.

#[ADDRESSING PLAN TEMPLATE]#

| Relative Prefix# | Subnet mask | Size | Addresses Used Immediate | 1yr | 2yr | Description |
|---|---|---|---|---|---|---|
| 0.0.0.0 | 255.255.255.128 | 128 | 80 | 95 | 100 | Department A |
| 0.0.0.128 | 255.255.255.224 | 32 | 12 | 17 | 25 | Department B |
| 0.0.0.160 | 255.255.255.224 | 32 | 0 | 15 | 28 | Department C |
| 0.0.0.192 | 255.255.255.240 | 16 | 7 | 8 | 10 | Department D |
| 0.0.0.208 | 255.255.255.240 | 16 | 10 | 14 | 14 | Department E |
| 0.0.0.224 | 255.255.255.240 | 16 | 11 | 12 | 12 | Department F |
|  |  | 240 | 120 | 161 | 189 | Totals |

An entry is made for each physical separate subnet in the network, subnets are considered to be physically separate if there is an IP router between them.

Prefix: or Relative Prefix: The network address of the subnet; if this Addressing Plan was being used to request addresses, then a relative prefix would be used where the first entry would be 0.0.0.0 and all subsequent entries would be sequential from that depending on the size of each subnet. For example, 0.0.0.0 with a mask of 255.255.255.0 would be followed by a prefix of 0.0.1.0, as shown above.
Subnet mask: The subnet mask of each entry.
Size: The size of the subnet based on the subnet mask.
Addresses Used: In these three fields, show the current and estimated usage for the next two years.
Current: The number of network interfaces currently used in the subnet.
Immediate: The number of addresses the organisation will use immediately.
1-yr: The number of addresses to be used in one year.
2-yr: The number of addresses to be used in two years.
Description: Used to specify a short description of the use of the subnet in the user's organisation.

**Fig 14.3** The 'Addressing Plan Template' is used to show how the organisation proposes to use the requested addresses

## 14.6 LIR's Address Space

How the assignments are made from the block of addresses allocated is left to the LIR to control. This allows some scope for network summarisation and simplification of the routing within the LIR's network. The ideal situation would be to allocate customers connected to each access node from specific ranges of addresses and maybe to have a large dial network from another block.

For example, say the LIR has the following block allocated to it — 194.72.0.0/16; this could be broken down:

194.72.0.0/18 = Customers in the north
194.72.64.0/18 = Customers in the midlands
194.72.128.0/18 = Customers in the south
194.72.192.0/19 = Dial network
194.72.224.0/20 = Infrastructure addresses
194.72.240.0/20 = SPARE

Every customer assignment made to customers connected to the LIR's network in the south would therefore come from the 194.72.128.0/18 block.

However, the blocks will not be used evenly and eventually one will run out or there will no longer be a large enough free range of addresses to satisfy an assignment. The spare addresses will therefore have to be used for the region which has run out of addresses:

194.72.0.0/18 = Customers in the north

194.72.64.0/18 = Customers in the midlands

194.72.128.0/18 = Customers in the south

194.72.192.0/19 = Dial network

194.72.224.0/20 = Infrastructure addresses

194.72.240.0/20 = More customers in the south

This is not a problem until one of the other blocks fills up and the LIR has to request more addresses from the RIR. In Europe, this is done by sending a list of all assignments from the allocation to RIPE who will audit them and ask for further details of some of them as evidence that the LIR is maintaining the address space properly. RIPE will not entertain a request for address space until 80% of the current allocation is used which in our example can cause further problems if one of the blocks is under-utilised. In order to fully use the block, the LIR may choose to spread one block over two regions:

194.72.0.0/18 = Customers in the north

194.72.64.0/18 = Customers in the midlands and some customers from the north

194.72.128.0/18 = Customers in the south

194.72.192.0/19 = Dial network

194.72.224.0/20 = Infrastructure addresses

194.72.240.0/20 = More customers in the south

As this progresses the advantages of network summarisation begin to erode; when the LIR is then allocated a further block of addresses, the process will continue. While it is not easy to strictly maintain the ideal model of address management across the network it should be considered as the network grows.

## 14.7 Address Management Tools

In order to keep track of their address space LIRs need to keep a top-level record of how their allocations have been split up — this is needed in the event of an audit by the RIR. A second layer of information is required which consists of detailed information about the assignments from the blocks of addresses, that have been used. Returning to the previous example the top-level record would appear something like:

228  *Address Management Tools*

194.72.0.0 Allocation 1
 194.72.0.0/18 Block for customers in the north
 194.72.64.0/18 Block for customers in the midlands
 194.72.128.0/18 Block for customers in the south
 194.72.192.0/19 Assigned to the dial network on RIR ticket number NCC#1234567
 194.72.224.0/20 Block for infrastructure addresses on RIR ticket number NCC#1234567
 194.72.240.0/20 Block 2 for customers in the south

The 194.72.192.0/19 range of addresses is directly assigned to the dial network and therefore does not have any underlying records of addresses — all of the other blocks contain underlying records:

194.72.0.0 Allocation 1
 194.72.0.0/18 Block for customers in the north
  194.72.0.0/25 Manchester Metalworkers Co RIR ticket number NCC#123444
  194.72.0.128/25 Scottish Tartan Ltd RIR ticket number NCC#127811
  etc
 194.72.64.0/18 Block for customers in the Midlands
  194.72.0.0/27 Birmingham Tyres
  194.72.1.0/24 Shakespeare Hotels  RIR ticket number NCC#122222
  194.72.3.0/29 Little Company
  etc
 194.72.128.0/18 Block for customers in the south
  194.72.0.0/23 London Bridge Co RIR ticket number NCC#124545
  194.72.3.0/24 Tower Bridge Co RIR ticket number NCC#126544
  194.72.22.0/28 Brighton Beach
  etc
 194.72.192.0/19 Assigned to the dial network on RIR ticket number NCC#1234567
 194.72.224.0/20 Block for infrastructure addresses on RIR ticket number NCC#1234999
  194.72.224.0/24 Block for connecting customers in the North
   194.72.224.1 Northern customers access router
   194.72.224.2 First customer on the northern customers access router
   194.72.224.254 Last customer on the northern customers access router
  194.72.225.0/24 Block for connecting customers in the midlands
   194.72.225.1 Midland customers access router
   194.72.225.2 First customer on the midland customers access router
   194.72.225.254 Last customer on the midland customers access router
  194.72.225.0/24 Block for connecting customers in the south
   194.72.225.1 Midland customers access router
   194.72.225.2 First customer on the southern customers access router
   194.72.225.254 Last customer on the southern customers access router

    194.72.226.0/24 Block for special customers connections
        194.72.226.0/30 Very fast customer
        194.72.226.4/30 BGP customer
        etc
    194.72.240.0 /20 Block 2 for customers in the south
    194.72.240.0/27 South coast hotels

Apart from the structure of the records there are a few things to note about the above records. Only the larger blocks have RIR assignment ticket numbers associated with them; this is because they are larger than the LIR's assignment window and have therefore had to have passed through the RIR's approval process. The assignments from within the blocks are not sequential — there is no requirement for this from the RIR. The assignment of the infrastructure top-level block has been approved by the RIR not the assignments from it — each sub-block of the infrastructure block would have been represented as a separate line in the infrastructure requests addressing plan.

An up-to-date copy of the addressing plan for the LIR's network infrastructure and other addresses used by the organisation, with current utilisation levels, should also be kept for when the organisation to which the LIR belongs requires more addresses. In the above example, the dial network addresses would fall into this category.

While there are a few commercial and freeware IP address record applications for aggregating address space, the maintenance of top-level records and addressing plans can adequately be carried out using spreadsheets.

All assignments and associated records should be registered in the RIR's public database and the tool for querying it is called 'Whois'. It can be accessed via Web pages [7] or it can be downloaded as an application and run on a local machine.

## 14.8  Address Conservation

A number of methods have been employed in order to extend the use of IP version 4 address space, two of which are examined below.

### 14.8.1  Network Adress Translation

Using software commonly available in routers and firewalls, it is possible to effectively hide whole networks using private addresses behind one or a pool of public addresses. There are two forms of network address translation (NAT). Port address translation (PAT), which is also known as 'NAT overload', works by utilising the port numbers available in the TCP/IP packet, and allocating a different port on the outside address to an internal address. Port numbers are represented by a

16-bit field in the IP packet header giving a total of 65 536 different ports available; however, many of these are reserved for use by specific applications. The number of inside addresses, which can be mapped to a single outside address, is reduced to a few thousand depending on the manufacturer's software. Extensive use of this method could go a long way towards the conservation of IPv4 address space. Unfortunately not all applications will work through PAT; for example, an address for an e-mail server which requires a globally unique address will be sent to the single outside address. To overcome this the more basic form of NAT allows a one-to-one mapping of an inside address to an outside address allowing the implementation of these services.

### 14.8.2 Dynamic Allocation of Addresses

Another method of saving public addresses is to dynamically allocate an address to the end user for the duration of their session connected to the network. This method is extensively used by an ISP's dial-up customers and allows a large number of users to use a relatively small pool of address space. Another example of this is dynamic host configuration protocol (DHCP), which allows end-user hosts to be configured automatically with an address for the duration of their use. When a DHCP client starts up it searches for a DHCP server and obtains set-up information, including an IP address; this can be used in combination with NAT and has the advantage of allowing changes to network configuration, which affect all clients, to be carried out centrally at the server.

## 14.9 Futures

Given the explosive growth of the Internet it is conceivable that a global IP address shortage could become a reality. Work is well under way towards the deployment of the next generation of the Internet based on IP version 6. IPv6 extends the size of addresses to 128 bits or $2^{128}$ addresses which equates to 340,282,366,920,938, 463,463,374,607,431,768,211,456 addresses. IP version 6 also adds a number of features including enhanced methods of dynamic address allocation, class-of-service-based routing and enhanced security features. The problem this creates is when and how to migrate or whether the existing addresses can be extended indefinitely.

## 14.10 Conclusions

This chapter has shown some of the issues BT has to face in order to successfully manage its vitally important IP address space within the overall global framework.

# IP Address Management 231

The consequences for getting this activity wrong could adversely affect BT's ability to operate in the new global IP environment.

# Appendix

## An example of the RIPE-141 form

Explanatory comments are shown in italic. The majority of the fields can be filled in from the information provided by the end user.

X-NCC-RegID: *each LIR has a unique RegistryID, this is used to identify who has sent the RIPE141 form in*

I. General Information
#[OVERVIEW OF ORGANIZATION TEMPLATE]#
*As provided by the end user and modified or added to by the person from the LIR making the request*

#[REQUESTER TEMPLATE]#
*Details of the person from the LIR making the request*

name:
organisation:
country:
phone:
fax-no:(optional)
e-mail:

#[USER TEMPLATE]#
*Details of the person from the end users organisation requesting the addresses*

name:
organisation:
country:
phone:
fax-no: (optional)
e-mail:

#[CURRENT ADDRESS SPACE USAGE TEMPLATE]#
*Details of all current public address usage by the end user organisation — see the Addressing Plan Diagram for further details*

|  |  |  | Addresses Used |  |  |  |
|---|---|---|---|---|---|---|
| Prefix# | Subnet mask | Size | Immediate | 1yr | 2yr | Description |

II. The Request
#[ REQUEST OVERVIEW TEMPLATE]#
Summary information about the request
request-size:             *information from the request addressing plan*
addresses-immediate:      *information from the request addressing plan*
addresses-year-1:         *information from the request addressing plan*
addresses-year-2:         *information from the request addressing plan*
subnets-immediate:        *information from the request addressing plan*
subnets-year-1:           *information from the request addressing plan*
subnets-year-2:           *information from the request addressing plan*

inet-connect: whether the organisation is currently connected to the Internet and if so how
country-net: the ISO 3166 two letter country code where the addresses will be used
private-considered: whether the organisation has considered the use of private addresses for the requested networks
request-refused: whether the organisation has ever been refused an IP address request in the past and if so who by and why
PI-requested: if the request is for Provider Independent address space
address-space-returned: whether the organisation is planning to return address space as part of this request

#[ ADDRESSING PLAN TEMPLATE]#
Details of the proposed use of the requested addresses - see the Addressing Plan Diagram for further details

|  |  | Addresses Used |  |  |  |  |
| --- | --- | --- | --- | --- | --- | --- |
| Prefix# | Subnet mask | Size | Immediate | 1yr | 2yr | Description |

III. Database Information
#[ NETWORK TEMPLATE ]#
Information that will be entered in the RIPE database about the organisation requesting the address space

inetnum:
netname:
descr:
descr:
country:
admin-c:
tech-c:
status:
mnt-by: (optional)
notify: (optional)
changed:
source: RIPE

#[ PERSON TEMPLATE ]#
If contact information for the network administration persons entered in the Network Template is already in the RIPE database then this should checked to see if it is up to date. For each person specified in the network template for which there is no entry in the RIPE one can be created using the following template

person:
address:
address:
address:
e-mail:
phone:
fax-no: (optional)
mnt-by: (optional)
notify: (optional)
nic-hdl:
changed:
source:
RIPE

#[TEMPLATES END]#
IV. Optional Information:

Any other information which may help the hostmaster with the request can be added here

## References

1  ARIN Web site — http://www.arin.net

2  RIPE Web site — http://www.ripe.net

3  APNIC Web site — http://www.apnic.net

4  AFRINIC Web site — http://www.afrinic.org

5  LATNIC Web site — http://www.latnic.org

6  ICANN Web site — http://www.icann.org

7  'Whois' tool — http://www.ripe.net/cgi-bin/whois

# 15

# TRAFFIC ENGINEERING

## S Spraggs

## 15.1 Introduction

Data traffic, particularly IP, continues to grow at an explosive pace. Leading Internet providers report bandwidths doubling on their backbones approximately every six to nine months. The Web-driven growth in demand for bandwidth of recent years will be followed by successive waves of demand attributable to voice over IP (VoIP), video, and high-speed subscriber access via digital subscriber lines (DSL) and cable. The adoption of intranets and extranets for networked commerce will bring further changes to the IP-service infrastructure, both through bandwidth demands and feature requirements.

Service providers recognise the unprecedented explosion of packet-based traffic and are re-evaluating their IP network architectures in the light of these changes. The traditional IP network, generally built as an overlay to a layer-2 technology such as frame relay or ATM, are being phased out in favour of simpler, more cost-effective models. These rely on the IP router connecting directly to dark fibre and/or dense 'wave division multiplexing' (WDM) equipment and effectively putting IP packets directly on to the glass. This scheme removes multiple layers of hierarchy and offers significantly more bandwidth, flexibility and cost savings but does offer some additional challenges.

One of the key challenges is how to utilise the network's resources to their best effect. In traditional IP networks, this was through a combination of the interior gateway protocol and the traffic engineering capabilities of the underlying layer-2 infrastructure. With the removal of the layer-2 infrastructure, this traffic engineering function needs to be performed elsewhere.

This chapter explains a technique called 'multiprotocol label switching (MPLS) traffic engineering (TE)' which operates alongside the IP layer to provide this functionality and looks at potential implementation models. TE is simply about utilising the circuits available in the network to best effect. MPLS TE has been included in this section of the book as MPLS is one of the more recent IP protocols to have an impact on the design and operation of carrier-scale IP networks. It should be noted that traffic engineering is not essential for the operation of all large IP networks — a careful cost benefit analysis should be conducted before deciding to operate MPLS TE on a particular network. Some networks may find MPLS TE too complex and expensive to implement compared to the costs of wasted bandwidth.

## 15.2 Problem Definition

In general, IP routing will place all packets on the shortest (or lowest cost) path to a destination regardless of congestion along that path or the availability of alternate unequal cost paths. This can lead to the so-called traffic concentration problem and result in significant network inefficiencies. Equal cost multipath forwarding (ECMF) enables flows to share equal cost paths, but this too can result in network inefficiencies due to the fact that one of the paths may be congested or more expensive than the cost metrics used to compute the path. Traffic engineering or the ability to compute and then 'steer' traffic flows on to non-default paths is a mechanism that solves this problem and one that is generating great interest in the ISP community.

Today the goal of traffic engineering is to maximise the utilisation of network resources. In a large network, available network bandwidth may not be efficiently utilised because, for each destination, an intra-domain routing protocol (e.g. OSPF, IS-IS) finds a single 'least-cost' route. However, this least-cost route may not be the only possible route, and it may not have enough resources to carry all the traffic. For example, in Fig 15.1, there are two paths between the San Francisco router and the New York router — San Francisco-Chicago-New York, and San Francisco-Dallas-Atlanta-New York. The routing protocol decides that the former path is preferred

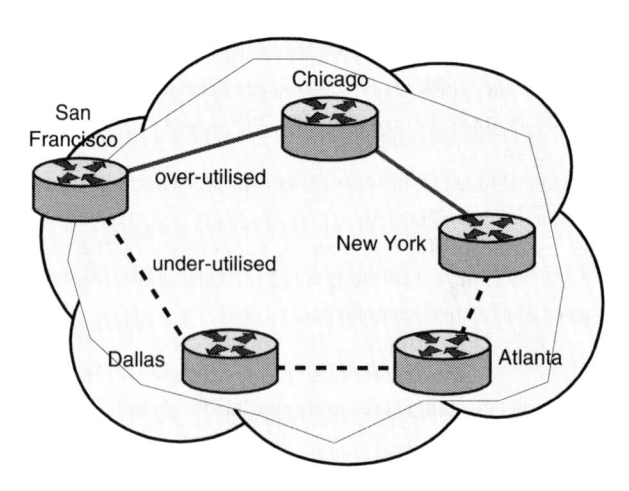

**Fig 15.1**  The traffic engineering problem.

and therefore all packets between these two points take the former path. Even when the San Francisco-Chicago-New York path is congested, packets are not routed to the San Francisco-Dallas-Atlanta-New York path, which is not congested.

Due to this limitation, the situation often arises where a part of the network is over-utilised while another part is under-utilised. Traffic engineering addresses this issue. In its current guise it is a capability primarily intended for the service provider to make better use of their underlying infrastructure, rather than a revenue-generating service offered to end-user customers.

## 15.3 MPLS TE Framework

RFC2702 outlines the requirements for traffic engineering over MPLS and identifies the main components and required functionality. The basic premise is unidirectional MPLS 'label-switch path traffic engineered tunnels' (LSP TE tunnels or TE tunnels) are built across the network into which IP traffic is routed. The route of a TE tunnel is determined by either static definition or a 'constraint based routing' calculation on the head-end router. A signalling mechanism then sets up the label switch path based on the previously calculated route. When an IP packet is routed into a TE tunnel it undergoes MPLS label imposition, which involves the placement of an MPLS label on to the IP packet. Once labelled, the path to the destination is determined by MPLS label switch forwarding rather than the IP prefix-based forwarding. The consequence is the route from tunnel head end to end point is determined by the path of the TE tunnel rather than the IP prefix, and hence offers the capabilities to move traffic off the interior gateway protocol's shortest path route.

In looking at an MPLS traffic engineering solution, it offers the following benefits:

- MPLS traffic engineering (TE) software enables an MPLS backbone to replicate and expand upon the traffic engineering capabilities of layer-2 ATM and frame relay networks;

- MPLS traffic engineering provides an integrated approach to traffic engineering — with MPLS, traffic engineering capabilities are integrated into layer 3, which optimises the routing of IP traffic, given the constraints imposed by backbone capacity and topology;

- MPLS traffic engineering routes traffic flows across a network based on the resources the traffic flow requires and the resources available in the network;

- MPLS traffic engineering employs 'constraint based routing (CBR),' in which the path for a traffic flow is the shortest path that meets the resource requirements (constraints) of the traffic flow — in MPLS traffic engineering, the flow has bandwidth requirements, media requirements, a priority versus other flows, etc;

- MPLS traffic engineering gracefully recovers to link or node failures that change the topology of the backbone by adapting to the new set of constraints.

### 15.3.1 Components of MPLS Traffic Engineering

MPLS TE is a collection of components and functions briefly described below. A fuller definition and explanation of the functionality and interaction between the various components are included in the subsequent sections:

- tunnel or traffic trunk attributes — these describe the network requirements for the trunk or tunnel;
- network resource attributes — these describe the traffic engineering attributes of the network, and are used as the input to the constraint based routing calculation;
- distribution of the network resource — the mechanisms employed to circulate the network resource attributes within the network;
- path selection — the process performed on the head-end router that determines the explicit path the TE tunnel will take when set up;
- TE tunnel set-up and maintenance — the set of processes used to signal, set up and maintain the label-forwarding tables on routers within the TE tunnel explicit path;
- routing traffic into TE tunnels — the way in which IP traffic can be mapped into a TE tunnel and how tunnels integrate into the IP routing topology.

It should be noted that all these functions occur prior to the forwarding of IP traffic across a TE tunnel.

### 15.3.2 Traffic Trunk Attributes

Traffic trunk attributes allow the network operator to describe the characteristics of traffic trunks or tunnel. They must be granular enough to account for the different types of packet traversing the network, and detailed enough to specify the desired behaviour in failure situations. There are six traffic trunk attributes and each is described below.

- Bandwidth

    This attribute specifies the amount of bandwidth the traffic trunk or tunnel requires.

- Path selection policy

    This attribute gives the network operator the option to specify the order in which the head-end routers should select explicit paths for traffic trunks. Explicit paths may be either administratively specified or dynamically computed.

- Resource class affinity

    This attribute is used to allow the network operator to apply path selection policies to administratively include or exclude network links. As will be shown later, each link on the network may be assigned a resource class as one of the resource attributes. Resource class affinity specifies whether to include or exclude links with resource classes in the path selection process. It takes the form of the tuple <resource class mask, resource affinity>. The 'resource class mask' attribute indicates which bits in the resource class need to be inspected, and the 'resource affinity' attribute indicates whether to explicitly include or explicitly exclude the links.

- Adaptability

    This attribute indicates whether the traffic trunk should be re-optimised. The re-optimisation procedure is discussed in a later section.

- Resilience

    This attribute specifies the desired behaviour under fault conditions, i.e. the path carrying the traffic trunk no longer exists due to either network failures or pre-emption. Restoration of MPLS TE tunnels is discussed in a later section.

- Priority

    Priority is the mechanism by which the operator controls access to resources when the resources are under contention. Another important application of the priority mechanism is supporting multiple classes of services. Two types of priority are assigned to each traffic trunk or tunnel — holding priority, and set-up priority. Holding priority determines whether the traffic trunk has the right to hold a resource reservation when other traffic trunks or tunnels attempt to take away its existing reservation. Set-up priority determines whether the traffic trunk has the right to take over the resources already reserved by other traffic trunks.

### 15.3.3 Network Resource Attributes

These are a set of attributes associated with network resources which constrain the placement of tunnels through them. They can be thought of as topology attribute constraints and are used as input to the 'constraint based routing' calculation performed by the head-end router to select a tunnel's path through the network. Each of the resource attributes is described below.

- Available bandwidth

    This attribute describes the amount of bandwidth available at each set-up priority. This attribute need not necessarily reflect the actual available bandwidth. In some cases, the network operator may oversubscribe a link by assigning a value which is larger than the actual bandwidth, e.g. 49.5 Mbit/s for a DS-3 link.

- Resource class

    This attribute indicates the resource class of a link. It should be recalled that the trunk attribute, resource class affinity, is used to allow the operator to administratively include or exclude links in path calculations. This capability is achieved by matching the resource class attribute of links with resource class affinity of traffic trunks. The resource class is a 32-bit value. The resource class affinity contains a 32-bit resource affinity attribute and an associated 32-bit resource class mask.

    A link cannot be considered in the path calculation for a traffic trunk unless the following equation holds:

    $$(\text{resource class}) \text{ \&} (\text{resource class mask}) = (\text{resource affinity})$$

    where '&' refers to a bitwise AND operation, and '=' refers to bitwise equality.

### 15.3.4 Distribution of Resource Attributes

The distribution of the resource attributes is a key requirement for traffic engineering if the route taken through the network of a TE tunnel is calculated using 'constraint based routing' mechanisms. Resource attributes are a router's view of the capabilities of its links but the concept of traffic engineering is about utilising the network, as a whole, in the most efficient fashion. Consequently, this information needs to be distributed to each router in the network so that it has an up-to-date view of the entire network and any existing constraints. In this way the source or head-end router of a TE tunnel can calculate the most effective explicit route for a TE tunnel, based on the tunnel's requirements and the current network topology.

The existing operation of link-state IGPs (IS-IS and OSPF) offers very similar capabilities. Each router in the network needs full knowledge of the network topology to calculate the shortest path table, from which the routing tables are derived. The difference is that existing IGP link state updates only contain the routing costs associated with each link, whereas for CBR to calculate a TE tunnel's explicit route the resource attributes are also required. Due to the similar requirements and operations, traffic engineering utilises existing link-state IGPs with extensions to distribute traffic engineering attributes within the network. Before discussing the specific implementation on OSPF and IS-IS, it is worth pointing out that there are a number of restrictions with traffic engineering when utilising 'constraint based routing' to calculate the explicit paths for TE tunnels. It should be noted that, if the tunnel path is statically configured on the head-end router, the restrictions listed below do not exist.

- Today traffic engineering only operates within a single area (in OSPF) or level (in IS-IS) when using constraint based routing to calculate the LSP path. The reason is that the full knowledge of the topology is hidden as routing information is moved between levels or areas — hence it is not possible for constraint based routing to see the necessary information to calculate the TE tunnel path.
- Distance vector protocols are not supported when using CBR to calculate the explicit route for the TE tunnel. This is because the head-end routers need a complete view of all possible paths between itself and the tunnel end-point to be able to select the most appropriate path. Distance vector protocols do not offer this, as they effectively supply the upstream router's view of the best pre-computed path.
- There is no support for CBR-calculated TE tunnels across 'autonomous system boundaries'. This is related to the above two points.

*15.3.4.1 OSPF*

The OSPF opaque LSA is used to transport resource attributes within the network. The opaque LSA is a generalised mechanism used to carry additional information within a network topology [1]. There are a number of different link-state types associated with the opaque LSA and are dependent on the flooding scope. The link-state-type field identifies the flooding scope — type 9 for link-local, type 10 for area-local and type 11 for AS. The current proposal for traffic engineering uses the type 10 area-local flood to distribute the LSA. The consequence of this is that within OSPF there is no capability today to use these mechanisms to distribute traffic engineering information across area boundaries. Through the extensions defined in IETF draft 'OSPF extensions for traffic engineering' [2], the opaque LSA is used to carry, among other things, the current and maximum bandwidth for each priority level and the other resource attributes described earlier.

*15.3.4.2 IS-IS*

New extensions have been designed and implemented for the IS-IS routing protocol to cater for traffic engineering. The extensions add additional IS-IS TLV (type, length, and value) objects that carry the same traffic engineering resource attributes as OSPF (for full details, refer to the IETF draft [3]).

*15.3.4.3 Flooding Details*

Transmission of resource attributes is a key consideration regardless of the routing protocol employed. Network resource attributes are both static and dynamic in nature. Attributes such as the affinity and TE administrative distance are essentially static while available bandwidth alters over time, based on the number and route of LSP tunnels within the network. Clearly this information needs to be flooded in the network and creates an interesting dilemma — too often and the routing protocol would threaten the overall operation of the network, too little and the head-end router's view of the network will be out of date. The consequence is that at tunnel set-up time the head-end router will compute a path to the tunnel destination based on out-of-date information. As a result, the RSVP path set-up may fail resulting in re-flooding of information, additional head-end calculations and additional RSVP signalling.

Within a network employing traffic engineering, normal link-state operation occurs except that the updates also contain traffic engineering attributes. This mechanism accounts for normal routing protocol operation and is able to deal with physical topology changes. To deal with specific changes in the traffic engineering attributes of the network, flooding also occurs when the resource class of a link changes. This can occur when the network operator reconfigures resource attributes and when the amount of available bandwidth crosses one of the preconfigured thresholds. The frequency of flooding is bounded by timers and also bandwidth 'up' and bandwidth 'down' thresholds. The 'up' thresholds are checked when a new TE tunnel traverses an MPLS label switch router (LSR), while the 'down' threshold is used when an existing tunnel goes away.

### 15.3.5 Path Selection

Path selection is the process used to define the hops to be taken from the source to the destination of the traffic trunk. There are two primary methods by which a TE tunnel can be set up, but regardless of the technique TE tunnels are effectively source routed across the network based on manual configuration or computation on the head-end router, or, put another way, where the tunnel was defined.

*15.3.5.1 Manual Configuration*

For each tunnel the network operator defines on the head-end router the destination and the intermediate hops from the source to the destination.

*15.3.5.2 Constraint Based Routing*

For each tunnel, the router starts from the destination of the trunk and attempts to find the shortest path towards the source (i.e. using the shortest path first (SPF) algorithm), but also taking into account the current capabilities of the network. This is done in two stages. The first stage is a pruning process, whereby links explicitly excluded by the resource-class affinities of the trunk and links which have insufficient bandwidth are removed from the topology. The second stage is a standard SPF calculation on the residual topology to calculate the shortest path to the destination.

### 15.3.6 Path Set-up, Admission Control and Path Maintenance

From the path-selection process (static or CBR-based) the head-end router now has an end-to-end view on how to reach the tunnel end-point. The path set-up procedure is initiated by the head-end router; its purpose is to confirm the network resources are still available on the devices and links within the explicit path and also to convey and set up the correct label information on all devices on the TE tunnel path. Two protocols have been proposed within the IETF for this label set-up procedure. They are 'constraint based routing — label distribution protocol' (CBR-LDP) and 'resource reservation protocol' (RSVP). This chapter's aims are practical in nature, in terms both of traffic engineering operation and of implementation in the real-world environment. Consequently it only looks at the RSVP implementation because at the time of writing it is the author's belief that only RSVP implementations are available.

When RSVP was originally conceived and defined in RFC2205 it was viewed as a 'resource reservation set-up protocol designed for an integrated services Internet'. Subsequently RSVP has been used for a number of other functions and it is now seen more as a generic signalling mechanism within an IP environment. One of the key areas where extensions have been proposed and implemented is in the use of RSVP to set up MPLS TE tunnels. These extensions and operations are covered in an IETF draft [4].

The RSVP-tunnel specification extends the original RSVP protocol by giving it new capabilities to support the set-up, maintenance and management of MPLS LSPs. This is achieved through the specification of several new RSVP objects and error messages that can be used in an MPLS domain.

Figure 15.2 illustrates the path set-up procedure for an explicitly routed TE tunnel and can be roughly summarised in the following steps (it should be noted that the explicit route may be based on either of the path-selection mechanisms described above).

- The head-end node creates an RSVP 'path' message containing LABEL-REQUEST object and the EXPLICIT-ROUTE object. The LABEL-REQUEST object indicates that a label binding is requested while the EXPLICIT-ROUTE object depicts the explicit route for the TE tunnel. Additionally the 'path' message also includes the bandwidth requirements and the other tunnel attributes of the TE tunnel.

244 *MPLS TE Framework*

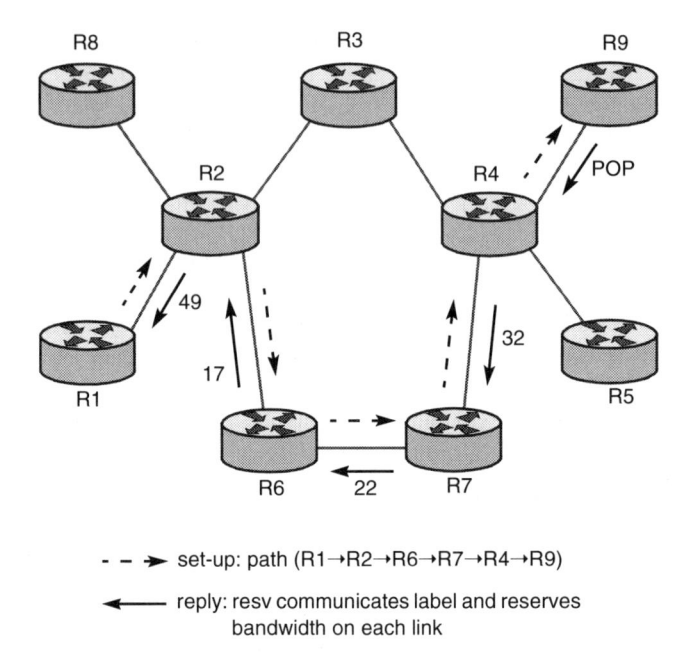

**Fig 15.2** Path set-up example.

- - ▶ set-up: path (R1→R2→R6→R7→R4→R9)

◀─── reply: resv communicates label and reserves bandwidth on each link

- When the 'path' message reaches the egress node of the LSP tunnel, a 'resv' message is created and a LABEL object containing an MPLS label is inserted into the 'resv' message. As the 'resv' message propagates to the origination node (in the reverse direction along the path traversed by the 'path' message), each node uses the MPLS label in the LABEL object from its downstream neighbour as the outgoing label for the LSP-tunnel. Each node inserts its own LABEL object before propagating the 'resv' message upstream. This way, labels are allocated sequentially all the way from the egress node of the LSP tunnel to the origination node.
- When the 'resv' message reaches the origination node, the LSP tunnel becomes established.

### 15.3.6.1 Admission Control

Admission control is performed on all intermediate routers upon receipt of the RSVP 'path' message. In principle, if the traffic engineering database on the head-end router is up-to-date, all the intermediate routers will have sufficient resource to honour the bandwidth requested within the path message. However, as previously discussed, it is not always practical to flood traffic attribute changes immediately; consequently there is a possibility the head-end router's view of the network does not reflect the actual state of the network. The router performing admission control uses bandwidth and tunnel-priority information in the 'path' message to check whether there is enough bandwidth to honour the reservation at the set-up priority of the tunnel. If there is enough bandwidth, the intermediate router creates 'path state' based on the tunnel's requirements and transmits the 'path' message on to the next router in the EXPLICIT ROUTE object. If the router performing admission control has insufficient resource to honour the request, an RSVP 'path error' message is generated and sent back to the tunnel head-end router. This 'path error' message travels the reverse path to the sender and removes RSVP 'path state' as it passes. If the 'path' message arrives at the destination of the TE tunnel, it generates the 'resv' message which is sent back to the tunnel head end and assigns labels and turns the 'path state' into a reservation. The trunk admission process on each individual router also has hooks into the routing protocol's flooding mechanisms. In the event of a reservation failure or if a link's bandwidth 'up' or 'down' threshold has been violated, due to a tunnel set-up or take-down, then the affected router floods this information via the routing protocol.

## 15.3.6.2 Re-optimisation

Re-optimisation is an important requirement of traffic engineering. It is the process whereby some traffic trunks are re-routed to new paths in order to improve the overall efficiency in bandwidth utilisation. Periodically, the network operator may need to modify the routing of traffic trunks because more conditions may have changed. The fluctuation in network topology may allow head-end routers of traffic trunks to find more optimal paths. For example, a traffic trunk may be re-routed to a secondary path due to a failure in the primary path. When the primary path is restored, it would be desirable to move the original traffic trunk back to the primary path. Without the re-optimisation capability, the network would become progressively sub-optimal every time a failure occurs. The adaptability trunk attribute indicates whether a traffic trunk can be re-routed for the reason of re-optimisation.

As mentioned above, there are two ways whereby re-optimisation takes place — via human intervention, or via automated procedures performed by the head-end routers. In the former case, the network operator must periodically update the traffic model stored in the head-end routers. In the latter case, the head-end routers must periodically recalculate paths for traffic trunks.

If a TE tunnel requires re-routing, it must be re-routed in a seamless fashion. With RSVP as the path set-up mechanism scheme, a process called 'make-before-break' is employed. The head-end router firstly establishes the new LSP for the traffic tunnel. When establishing the new LSP the head-end router indicates that the new reservation uses the shared-explicit mode. This way, the bandwidth is not reserved twice (known as double-counting) on the links that are shared by the old LSP and new LSP.

Once the new LSP is in place, then the traffic is switched over and the old path is torn down.

### 15.3.6.3 Restoration

Traffic engineering supports two complementary restoration techniques — path protection and link protection. Path protection is controlled at the head end of the tunnel. When the head-end router detects that the LSP carrying the tunnel has failed, it performs the action specified by the resilience attribute of the tunnel. The possible actions are:

- none — no explicit path protection which means that in failure scenarios the packets will be forwarded based on normal IGP routing;

- fallback to pre-computed, pre-established LSP — this option requires that a back-up LSP be pre-computed and pre-established, thus providing the fastest path restoration at the expense of wasting the bandwidth reserved by the back-up LSP;

- establish an LSP based on pre-computed path and move the trunk on to the new LSP — this option requires that the head-end router pre-compute a back-up path for the traffic trunk (since the LSP is established on-demand, there is no need to maintain idle LSPs); the price for the bandwidth efficiency over the previous option is the additional delay incurred for establishing the new LSP;

- compute and establish a new path, and move the trunk on to the new path — this option provides the slowest restoration but incurs no bandwidth or processing overhead.

The response time of path restoration is equal to the sum of the following delay components:

- the time it takes for the router closest to the failure to detect it;

- the time it takes for the head-end router to detect the failure (via either an explicit RSVP failure notification or IGP flooding);

- the time it takes to compute and establish a new path — note that if the fall-back path is pre-computed and/or pre-established, this delay component is partially or completely eliminated.

In some cases (e.g. voice over IP applications), this response time may be too long. Link protection is a faster restoration mechanism.

Link protection, as the name implies, aims at protecting individual links. Therefore, for each link, link protection is controlled by the routers at the two ends of the link. In the case of a link failure, the routers at the ends of the links switch the appropriate traffic trunks on to a back-up link. As in the case of path protection, this back-up link may be pre-established, or pre-computed but established in real time, or computed and established in real time. The resilience attribute of the traffic trunk is used to specify whether the trunk may utilise link protection. Link protection is supported using nested LSPs. Figure 15.3 demonstrates how link protection works.

In the example shown in Fig 15.3, the traffic trunk originates at R1 and terminates at R9. The link between R2 and R4 is protected using link protection. The normal path calculation yields the path R1-R2-R4-R9, and an associated LSP is created. A back-up path, R2-R6-R7-R4, is configured for the link between R2 and R4. An LSP is also established for the back-up path. If the link between R2 and R4 fails, R2 re-routes the traffic trunk to the back-up link by the following — instead of swapping label 37 to label 14 and forwarding the packet towards R4, R2 swaps 37 to 14 and also pushes label 17 on to the label stack and forwards the packet towards R6. R6 swaps label 17 with label 22 and forwards the packet towards R7. R7 pops the label stack once and sends the packet towards R4. As before, R4 receives the packet with label 14 and continues to forward the packet as in the normal case.

The concept of link protection may be applied to node protection. Node protection can be accomplished by configuring fall-back paths which route around individual nodes.

Link and node protection is triggered as soon as the router interface detects that a link failure has occurred. However, it is insufficient as the only restoration technique because:

- it does not take into account the efficiency of bandwidth utilisation;
- the back-up link may run idle when link protection is not triggered.

### 15.3.7 Routing Traffic into TE Tunnels

If all the processes previously described have executed error free, TE tunnels are then established across the network. The next challenge is to get traffic to utilise the TE tunnels. Before examining this area in more detail it is important to understand what a tunnel looks like to the routing process on the router. It appears as a unidirectional pipe that starts from the head-end router and terminates on the destination router. Only the head-end router can send traffic on to the tunnel and it appears to the head-end router as a logical interface. Therefore, providing the tunnel is in an active state, the head-end router can route traffic to the logical interface in exactly the same way as traffic can be routed on to a physical interface. It should be noted that intermediate routers have no way in which to get traffic directly on to the TE tunnel. That said, in some situations an intermediate router may route IP traffic to a tunnel head-end router that subsequently traverses that router again within a TE tunnel. This would be an unusual situation and would only occur in the event of static routing or where the TE tunnel's routing metric has been manipulated.

traffic trunk from R1 to R9
link protection between R2 and R4

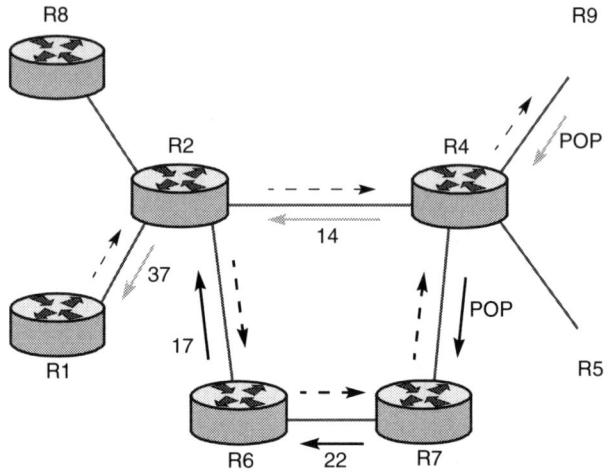

**Fig 15.3** Link protection example.

normal path: path (R1→R2→R4→R9), TSP (37, 14, POP)

link protection path: path (R1→R2→R6→R7→R4→R9), TSP (37, <17,14>, <22,14>, 14, POP)

As suggested, the key to getting traffic on to a TE tunnel is through the routing process and this can be achieved using static or an 'interior gateway protocol' such as IS-IS or OSPF. In the case of static routing the process and operation is fairly obvious. Utilising the IGP is more interesting as it offers a mechanism to build a process that is automatic in operation, but that also requires alterations to the operation of the IGP to cater for the TE tunnels.

### 15.3.7.1 Mapping Traffic into Tunnels Using Link-State Algorithms

Before discussing the mechanisms employed by the IGP to route traffic into TE tunnels, it is important to describe how the Dijkstra SPF algorithm is used in traffic-engineered environments. Firstly, it is for computing explicit routes for TE tunnels based on a pruned topology map using techniques associated with CBR. Secondly, it is used by the IGP to build the shortest path tree from which the routing tables are derived. This section concentrates on the latter of these two uses and looks at how the IGPs have been adapted to deal with tunnels and how traffic is routed on to these tunnels. It is heavily based on two IETF drafts [5, 6].

### 15.3.7.2 Enhancement to the SPF Computation

During each step of the SPF computation, a router discovers the path to one node in the network. If that node is directly connected to the calculating router, the first-hop information is derived from the adjacency database. If a node is not directly connected to the calculating router, the node inherits the first-hop information from the parent(s) of that node. Each node has one or more parents and each node is the parent of zero or more downstream nodes.

To deal with TE tunnels the SPF logic has been altered. It now considers TE tunnels as the possible first hop to a node when moved from the TENTative list to the PATHS list. With the introduction of the traffic engineering the SPF logic uses the following steps to determine the first-hop:

- examine the list of tail-end routers directly reachable by way of a TE tunnel — if there is a TE tunnel to this node, use the TE tunnel as the first-hop;
- if there is no TE tunnel, and the node is directly connected, use the first-hop information from the adjacency database;
- if the node is not directly connected, and is not directly reachable by way of a TE tunnel, the first-hop information is copied from the parent nodes to the new node.

As a result of this computation, traffic to nodes that are the tail end of TE tunnels flows over the TE tunnels. Traffic to nodes that are downstream of the tail-end nodes also flows over the TE tunnels. If there is more than one TE tunnel to different intermediate nodes on the path to destination node X, traffic flows over the TE tunnel whose tail-end node is closest to node X.

All other functionality such as equal-cost parallel paths to destinations does not change as a result of the enhancements and is able to deal with scenarios where the traffic is forwarded over native IP paths, TE tunnels or a combination of both.

### 15.3.7.3 Route Installation and Metrics

The completed path table is used to build the 'routing information base' (RIB). The default behaviour when computing the distance or metric for routes installed into the RIB over TE tunnels is to inherit the metric of the shortest native IP path. For example router A can reach router C with the shortest distance of 20. X is a route advertised in IGP by router C. Route X is installed in router A's routing table with the metric of 20. When a TE tunnel from router A to router C comes up, by default the route is installed with a metric of 20, but the next-hop information for X is changed to point to the tunnel.

In most situations this is quite satisfactory; however, there may be scenarios where further manipulation is required. In these cases it is possible to alter the metrics associated with a tunnel in an absolute or relative fashion. With absolute metrics no account is taken of the underlying IP topology and the metric is fixed and assigned by the network operator during TE tunnel configuration. Assuming the TE tunnel represents the preferred first-hop to the tunnel end-point, then the metrics of IP prefixes, advertised by the tunnel end-point and any downstream nodes, are related to those of the tunnel.

In relative mode the metric derived from the network topology is manipulated in either a positive or negative fashion. This manipulation of metrics does not impact the SPF algorithm. Its impact is whether the tunnel LSP is installed as one of the next hops to the destination routers and the metric value of routes installed into the RIB. The mechanism is also loop free because the traffic through the TE tunnels is basically source routed. The end result of TE tunnel metric adjustment is the control of traffic load-sharing. If there is only one way to reach the destination through a single LSP tunnel, then no matter what metric is assigned, the traffic has only one way to go.

## 15.4 Traffic Engineering QoS

Traffic engineering is a new discipline in MPLS environments and the first implementations do not offer quality of service capabilities directly. This statement is prone to misinterpretation and misunderstanding. With current traffic engineering it is not possible to set up a TE tunnel and direct all the high-priority traffic to a particular destination prefix down it, while sending lower and best-effort traffic to the same destination over a different path. It is currently possible though for traffic sent over an MPLS traffic engineered tunnel to utilise the underlying MPL 'diffserv' infrastructure based on 'experimental bit-inferred CoS' [7]. This is done by a combination of functions. When a packet is placed in a TE tunnel it undergoes label imposition.

This is the process of matching the IP destination address, establishing the next-hop address and imposing the MPLS label for that next-hop address. Part of the current label imposition process copies the IP precedence bits and places them in the EXP bits in the MPLS label (see Fig 15.4).

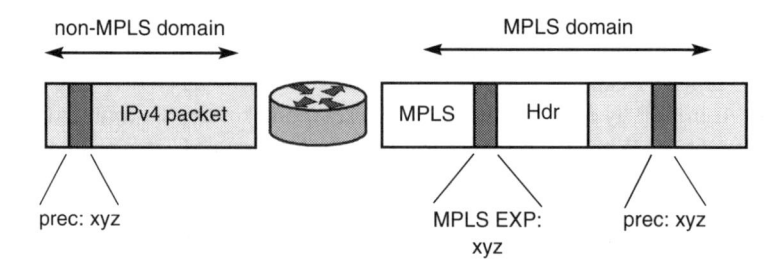

**Fig 15.4** IP precedence to MPLS EXP mapping.

If further label imposition occurs over the MPLS path the EXP bits are copied to all higher level labels, e.g when invoking the link protection mechanism described previously. Therefore if the IP network is using the IP precedence bits to convey QoS, these are reflected up the label stack in the EXP bits and the MPLS forwarding components can schedule and manage the queues based on these values. This mechanism provides up to eight different classes.

## 15.5 Traffic Engineering Configuration

Before traffic engineering can be used, the network needs to be prepared. There are two basic requirements. The first is to configure the tunnels and their attributes; this determines where the tunnels go and end, and also the characteristics of the tunnel. The second task is to configure the resource attributes of the links. These describe the capabilities of the links. Once this has been done, the traffic engineering can take place in the network.

## 15.6 Implementing Traffic Engineering

Although the role of traffic engineering in the network is clear, how it is used and implemented is dependent upon the exact configuration and the specific problem to be solved.

### 15.6.1 By Exception

In this mode of operation 'traffic engineering' is used sparingly on a 'by exception basis'. Normally traffic is routed over the IGP shortest path route. If the shortest path routing is creating bottle-necks or hot-spots within the network, based either on topology or traffic patterns, then traffic engineering is utilised to alleviate these situations. Figures 15.5 and 15.6 show two cases where this mode of operation could be used to great benefit.

Figure 15.5 shows an example where a particular host site is highly utilised. This may be due to on-going popularity, it is new, it has moved or it is attracting very high levels of short-term interest. A good example of the last is the Web site that first published the verdict in the Monica Lewinsky trial. In all the above cases the network may experience congestion on certain links or routers. In the case of on-going popularity, it could legitimately be argued that normal day-to-day capacity planning should deal with this scenario. However, all the other cases it is extremely difficult to control or predict, let alone get new capacity to deal with transient conditions within the network. A potential solution is to define an explicit tunnel that does not utilise the shortest path route and send some traffic down this tunnel using static routes as shown in Fig 15.5.

252 *Implementing Traffic Engineering*

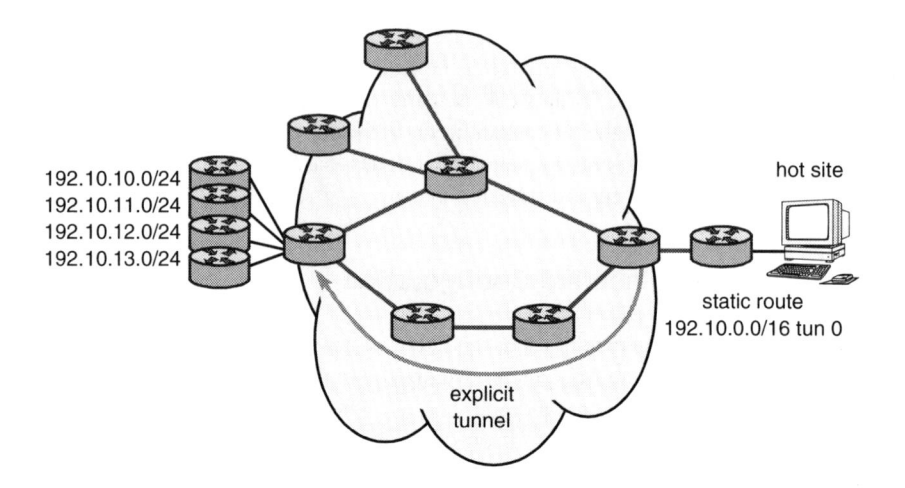

**Fig 15.5**  The hot site.

Figure 15.6 shows an example where the routers will see external routing information arriving via two BGP-4 peering connections. When BGP is faced with multiple routes to the same destination, it chooses the best route for routing traffic towards the destination. The following process summarises default BGP behaviour:

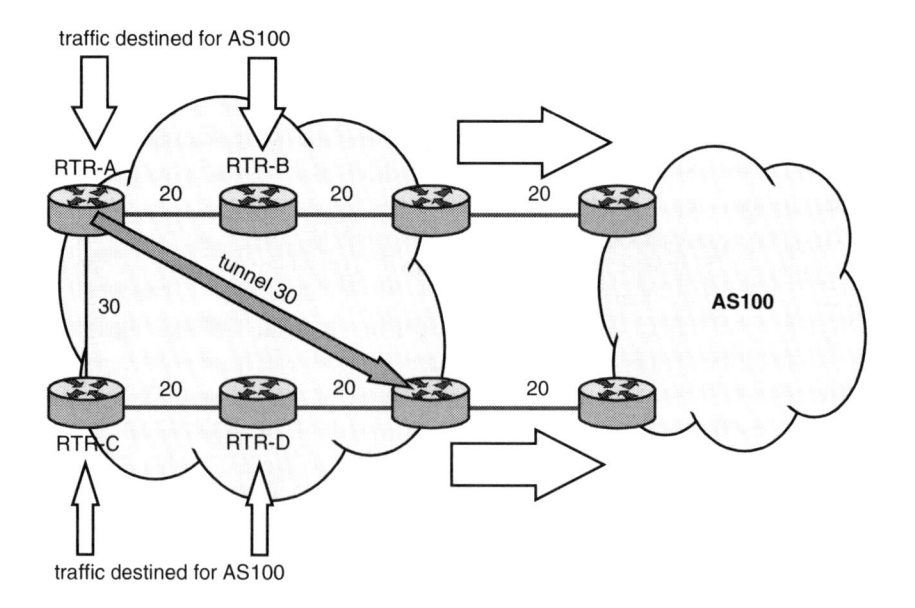

**Fig 15.6**  Unbalanced exit point.

- if next hop is inaccessible the route is ignored;
- prefer the path with the largest weight;
- if weight is the same, prefer path with largest local preference;
- if routes have the same local preference, prefer the route that was locally originated;
- if local preference is the same, prefer the route with the shortest AS_PATH;
- if AS_PATH is the same, prefer the route with the lowest origin type;
- if the origin type is the same, prefer route with the lowest MED;
- if the routes have the same MED, prefer the eBGP over confederation eBGP over iBGP;
- if all preceding scenarios are the same, prefer the route that can be reached via the closest IGP neighbour, i.e. the shortest internal path to the AS to reach the destination — in other words follow the shortest path to the BGP NEXT_HOP;
- if internal path is the same, a tie-break occurs based on BGP router ID.

Assuming the decision criteria to leave the network on the left to AS100 is based on the shortest internal path to the BGP NEXT_HOP address, then a TE tunnel can influence the exit point into AS100. Prior to defining the tunnel, the traffic loads, represented by the width of the arrows, shows RTR-A and RTR-B are receiving substantially more traffic destined for AS100 than the RTR-C and RTR-D. As a consequence, the top BGP peering router is heavily loaded while the bottom is only lightly loaded. By using traffic engineering the dynamics of the network are altered. In this case, a TE tunnel is set up, either explicitly or using constraint based routing, from RTR-A to the bottom BGP peering router. The tunnel is announced to the IGP for inclusion in the IGP routing calculations. By default, the IGP sees the tunnel as a connection with a metric equal to the IGP's lowest cost path to the tunnel end-point. Therefore, in Fig 15.6, the tunnel would have a cost of 90 to the BGP NEXT_HOP. As previously discussed traffic engineering has the capabilities to override the IGP metric. In the example the metric of the tunnel is reduced to 30. Now RTR-A sees the BGP NEXT_HOP of the bottom peering router with a cost of 50 and the BGP NEXT_HOP of the top peering router with a cost of 60. As a consequence the BGP routes associated with the bottom peering router will be installed in the routing tables of RTR-A.

The examples illustrate how traffic engineering on an exception basis can aid the capacity planners and the network designers in dealing with either short-term transient conditions or long-term imbalances caused either by traffic profiles or constraints imposed by the shortest path routing. As the examples illustrate, this mode of operation tends to rely on manual configuration. Typically tunnel paths are configured explicitly and traffic is directed on to the tunnels by either static routes or manipulation of the IGP metrics.

254 *Implementing Traffic Engineering*

To use this type of traffic engineering it is necessary to have a very clear view of the link utilisation in the network and traffic matrix statistics. Once these two pieces of information have been collated, it is possible, using TE tunnels, to steer traffic based on prefix or exit point away from network hotspots. Monitoring and management of this environment is complex with the need to consider the IGP derived shortest path routes, explicitly configured tunnels and diverting specific prefixes on to the tunnels via static routes.

### 15.6.2 Meshed Tunnels

This mode of operation calls for a mesh of LSP tunnels between routers within the network. The aim is to utilise traffic engineering to make the best use of all available assets within the network.

Figure 15.7 shows a very simple network. As a whole this network is not used efficiently with shortest path routing. With shortest path routing the link between RTR-B and RTR-C is not used during normal operation. By implementing a mesh of LSP tunnels the network as a whole is better utilised. Using RTR-C as an example, shortest path routing gives:

- to RTR-B — C-A-B is cost 20, C-B is cost 30, therefore C-A-B is selected;

- to RTR-A — C-A is cost 10, C-B-A is cost 40, therefore C-A is selected;

- RTR-C to ETR-A — traffic of 80 + RTR-C to RTR-B, traffic of 25 > CA link bandwidth of 100, therefore congestion is experienced;

- RTR-C to RTR-B — no traffic will traverse this link..

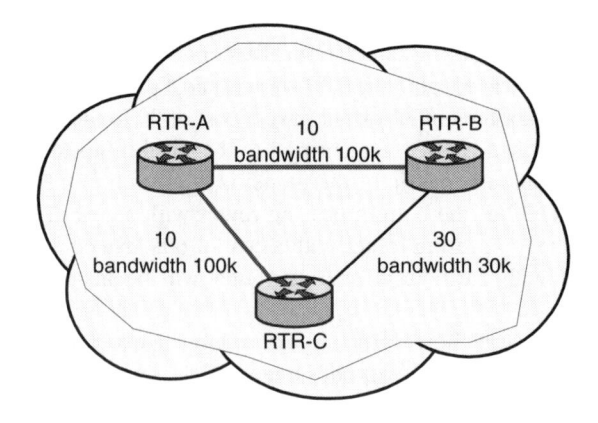

**Fig 15.7** Use of meshed environment.

By configuring tunnels from RTR-C and RTR-B and allowing the IGP to use the tunnels, the network as a whole is better utilised. The tunnel to RTR-A would go over the 100 kbit/s link. The TE tunnel to RTR-B would not be able to use the 100 kbit/s line, as there is not enough available bandwidth for an 80 kbit/s tunnel and a 25 kbit/s tunnel. Consequently the tunnel to RTR-B would set up over the 30 kbit/s link. The IGP now routes prefixes associated with RTR-A over the tunnel 0 and prefixes RTR-B over tunnel 1. Clearly this is a contrived example but it illustrates the potential applications and benefits of a meshed tunnel infrastructure

In a meshed environment TE tunnels are typically configured to set up dynamically based on CBR and to utilise the underlying IGP to determine which prefixes flow over the tunnels. To use the mesh approach effectively a very clear view of the aggregate bandwidth requirements between each end-point, along with the current loading of the actual underlying infrastructure, is needed. Although useful, it is not strictly necessary to have as granular a view of the network traffic matrix as in the 'by exception' TE model. This is because all traffic in the mesh is running over the TE tunnels. Therefore, provided the tunnel statistics are monitored, the tunnel bandwidth configuration is adjusted and the underlying bandwidth is confirmed as being sufficient, the tunnel set-up, tunnel maintenance mechanisms and the IGP can be relied upon to optimise the network. In some scenarios traffic engineering, in a meshed environment, can be further enhanced with multiple tunnels between each end-point. Clearly more configuration is required, but the benefit is that traffic can be load balanced across the tunnels and links (see section 15.6.3).

It is worth mentioning that, although the network appears to be meshed, the underlying IGP continues to operate over the physical network infrastructure and routing adjacencies are not formed over the tunnels. This means a TE tunnel mesh does not experience the same IGP flooding issues often seen in a full mesh ATM or frame relay routed network.

A key decision in this model is where the edge of the mesh resides. This depends on the physical infrastructure and why 'traffic engineering' is being used. A number of options are presented below:

- Core mesh

    At the edge of the traffic engineered environment are the routers that connect the PoP infrastructure to the core WAN links. In this way all traffic travelling between PoP sites is traffic engineered, and the WAN network can therefore be optimised based on overall traffic requirements of the network. Within the PoP itself normal IP routing mechanisms are employed. In most cases this will be sufficient, as generally the bandwidth within the PoP is an order of magnitude greater than the WAN environment and it is normally considerably cheaper to provision (Fig 15.8).

256 *Implementing Traffic Engineering*

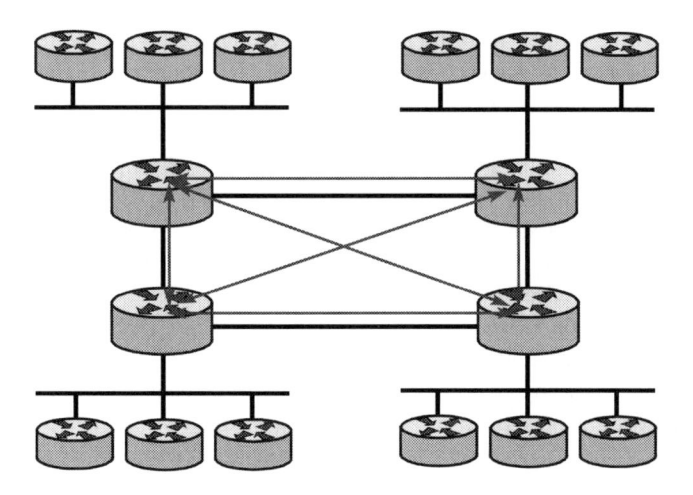

**Fig 15.8** Core mesh.

- Full mesh

    In this case the whole network is meshed such that all edge routers are fully meshed. This allows end-to-end traffic engineering from ingress to egress of the network.

    In some situations, such as with slow access devices connected via slow links to the core, this may be useful. The downside is that, in a large network, it represents a major overhead in tunnel configuration because, as tunnels are unidirectional, there needs to be $n(n-1)$ tunnels (where $n$ is the number of nodes in the mesh) (see Fig 15.9).

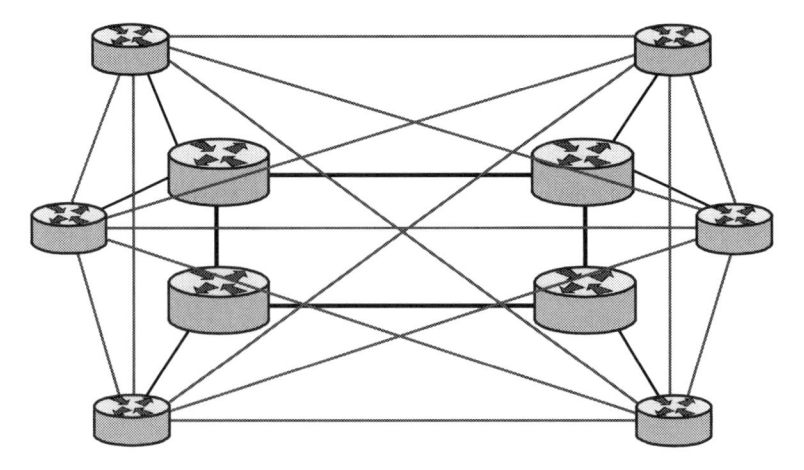

**Fig 15.9** Full mesh.

- Hierarchical meshing

  This achieves roughly the same affect as a fully meshed network but in a hierarchical fashion with significantly less configuration and offers considerably more flexibility (see Fig 15.10).

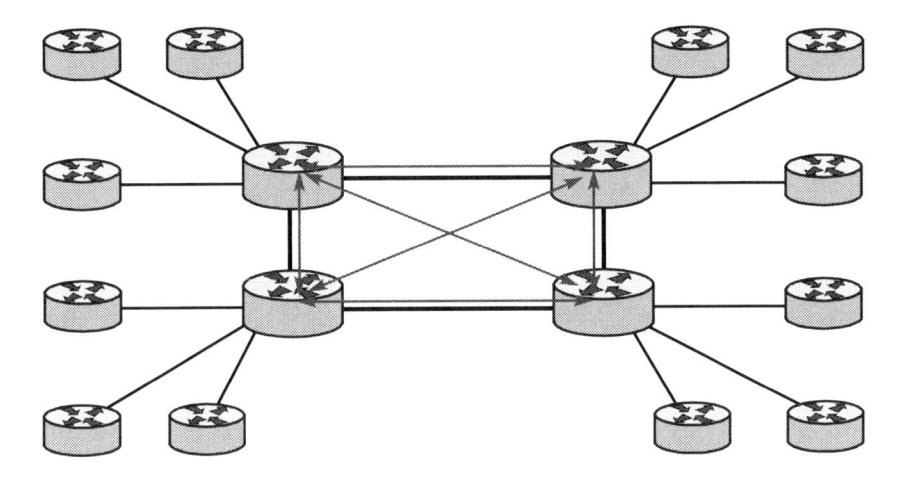

**Fig 15.10** Hierarchical mesh.

- Partial meshing

  This is a half-way step between a meshed environment and a 'by exception' environment. The concept uses IGP shortest path routing in the network except in key areas where there is a specific need for traffic engineering capabilities. In these areas the TE tunnels are configured either bidirectionally or unidirectionally. A good example might be where there are a number of very expensive, comparatively low-bandwidth links where it is important to get the best utilisation out of these facilities. In this case the area around the links would be traffic engineered in such a way that the only path on to the links is via a TE tunnel, while on either side normal shortest path routing is used (see Fig 15.11).

  In deciding the approach the following considerations should be looked at.

- Numbers of tunnels

  Meshing access to access throughout the network is going to require high levels of configuration and ongoing configuration when any kind of new access router is brought into the network. By meshing to the PoP core routers the number of tunnels needed is significantly less than a mesh down to the access layer. Furthermore, as the core PoP routers tend to be more static in nature, the level of on-going configuration requirements are less.

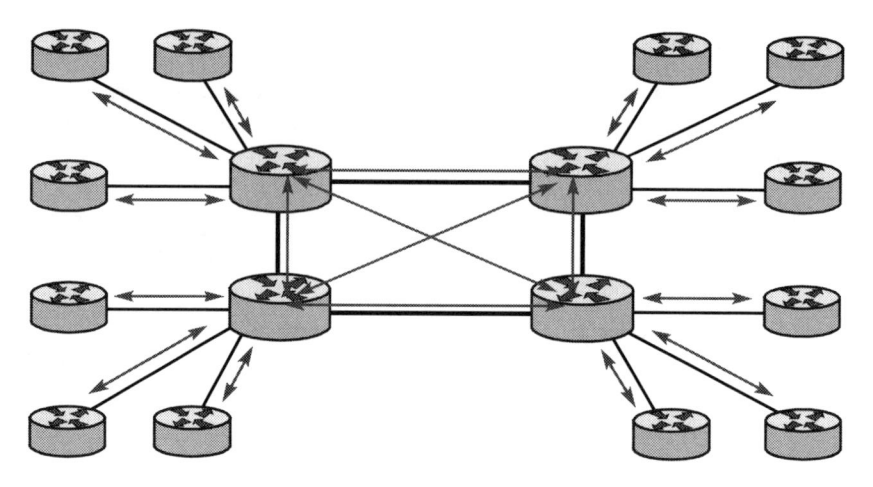

**Fig 15.11** Partial mesh.

- Purpose of tunnels

    Are tunnels really needed in the intra-PoP environment? Traffic engineering is currently designed to make best use of available bandwidth. Typically this is not a problem within the PoP because the infrastructure normally has plenty of bandwidth and, if additional bandwidth is required, this tends to be cheap and fairly easy to install. The need for traffic engineering to the access layer will become more important as the use of tunnels expands, e.g. if a tunnel is built to provide a guaranteed bandwidth connection.

### 15.6.3 Load-Balanced Self-Regulating Tunnels

TE tunnels have a number of unique capabilities that can maintain tight control of the paths across the network and also have two levels of load balancing based both on the IGP view of the world and also on the relative bandwidth of the configured tunnels. This combination of facilities means another use of traffic engineering is to build a network, or parts of the network, that use all available links, regardless of the topology, between two points. Through manipulation of tunnel parameters it is possible to load balance over these different links, based either on artificial mechanisms or on the relative bandwidth of the physical infrastructure. This application of traffic engineering is very useful where there is little control over the source or the destination of the traffic and where particular links are very expensive. A good example is a European or Asian service provider with presence in the USA and multiple connections across the ocean.

This is well illustrated by Fig 15.12. On the assumption that the left-hand side of Fig 15.12 is the USA and the right is Europe and the vast majority of the traffic is coming from net 1 destined for net 5, with traditional IP routing it is extremely difficult to get both links to be utilised or indeed the bottom one used at all. TE tunnels deal with this extremely well. In this case two TE tunnels would be set up, one over each of the two potential paths. From a routing perspective, R1 would now see net 5 via the two tunnel interfaces. If no bandwidth parameters are specified on the tunnel interfaces or they are equal, the traffic destined for net 5 would now be load balanced equally (based on IGP equal-cost load balancing) across the two tunnels, and hence links. If further control of link usage is needed, this can be done using the tunnel bandwidth parameter during configuration. This allows traffic to be load balanced, based on 'Cisco express forwarding' (CEF), but in relation to the bandwidth specified on the tunnels. For example, if the top tunnel has a bandwidth of 200 and the bottom tunnel 100, the traffic would be shared, based on a 2:1 ratio.

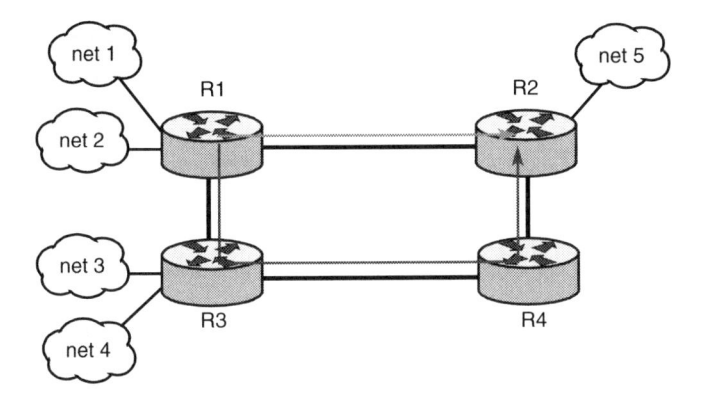

**Fig 15.12**  Load-balanced, self-regulating traffic engineering.

Using these mechanisms allows expensive link capacity to be used in a predefined ratio with little regard for the traffic mix or changes in traffic profiles. To complete the above picture, two tunnels would also be configured, over different paths, from each US router to each European router.

Management is comparatively simple as the TE tunnels load balance in a predefined way, regardless of traffic source or destination and loads. Therefore monitoring the utilisation of physical links is all that is strictly required.

## 15.7   Conclusions

As IP networks become larger and represent a significant portion of an organisation's traffic, so the trend in IP network design is towards solutions that transport IP efficiently and with minimal layers of abstraction between the physical transmission equipment and the IP routers. Recent advances, such as 'Packet over

SONET/SDH' and the ability for routers to connect either directly to fibre or DWDM systems, make IP-over-glass a reality and indeed a necessity to compete effectively in the IP world. These advances have exposed some weaknesses in traditional interior gateway protocols and their path selection mechanisms. The aim of the first implementations of MPLS traffic engineering is not only to allow IP networks to be built efficiently but also to alleviate the shortcomings with shortest path routing, such that the overall network infrastructure is used to its best advantage. In the longer term it is anticipated that MPLS traffic engineering will offer the capability to build dedicated paths through the network for different traffic types (e.g. QoS classes).

## References

1 Coltun, R.: '*RFC 2370 — The OSPF Opaque LSA option*', (July 1998).

2 Yeung, D.: '*OSPF extensions for traffic engineering*', (draft-yeung-ospf-traffic-00.txt).

3 Smit, H. and Li, T.: '*IS-IS extensions for traffic engineering*', (draft-ietf-isis-traffic-01.txt).

4 Awduche, D., Berger, L., Gan, D. H., Li, T., Swallow, G. and Srinivasan, V.: '*Extensions to RSVP for TE tunnels*', (draft-ietf-mpls-rsvp-lsp-tunnel-03.txt).

5 Li, T., Swallow, G. and Awduche, D.: '*IGP requirements for traffic engineering with MPLS*', (draft-li-mpls-igp-te-00.txt).

6 Shen, N. and Smit, H.: '*Calculating IGP route over traffic engineering tunnels*', (draft-hsmit-mpls-igp-spf-00.txt).

7 Davie, B., Heinanen, J., Vaananen, P., Wu, L., Le, Faucheur F., Cheval, P., Krishnan, R. and Davari, S.: '*MPLS support of differentiated services*', (draft-ietf-mpls-diff-ext-03.txt).

# 16

# IP VIRTUAL PRIVATE NETWORKS

## S Hills, D McGlaughlin and N Hanafi

## 16.1 Introduction

Carrier-scale IP networks can offer more than just Internet access, they can be used to deliver virtual private networks (VPNs), a service traditionally offered by frame relay and ATM networks. A VPN uses a shared infrastructure to carry traffic for multiple domains (e.g. different customers or communities of interest). Privacy (i.e. traffic separation) is provided using various techniques that can reside at layer 2 or layer 3 of the OSI model. IP VPNs, in particular, apply segregation at layer 3.

The key factor in VPNs is that all traffic from one domain (or customer) shares the same infrastructure as other domains, leading to economies of scale. This is achieved while maintaining security and separation from other domains (or customers). The main driver to the development of IP VPNs is therefore cost.

This chapter concentrates on the latest IP VPN technologies — tag switching (MPLS, see Chapter 15) and IP security (IPSec) — compares them to legacy systems, and positions them against each other.

## 16.2 Tag Switched VPNs

### 16.2.1 What is a Tag Switched VPN?

Tag switching was developed to advance the work of the Multiprotocol Label Switching (MPLS) Working Group in the IETF. Cisco's tag switching [1] uses MPLS [2] as its basis but provides pre-standard features, such as provider-based tag VPNs. BT's and Concert's tag switched network is built to offer a managed IP VPN service using Cisco routers. Subsequently the term 'tag' is used in preference to MPLS throughout this chapter. Likewise this chapter focuses primarily on features associated with Cisco routers when used in the tag VPN environment.

Tag switching can be used in the implementation of VPNs, providing customers with a dedicated private network over a shared IP infrastructure. Isolation between each customer's traffic is achieved by marking it with a separate unique routing identifier per VPN, allowing each VPN to maintain its own separate routing tables and functions [2].

Tag switching is a proprietary protocol from Cisco that combines the benefits of network layer (layer 3) routing and data link layer (layer 2) switching to provide scalable, high-speed switching in a shared network infrastructure. Tag switching technology is based on the concept of label swapping, whereby data packets are assigned short, fixed-length labels that tell tag switching nodes how data should be forwarded. As each tag is of a fixed length, look-up and forwarding through the tag switched network are fast and simple.

Tag information can be carried in a packet in a variety of ways, for instance:

- the tag can be inserted between the layer-2 and network-layer headers;

- the tag can be inserted as a part of the layer-2 header if the layer-2 header provides adequate semantics (such as ATM).

This flexibility in the positioning of the tag enables tag switching to be implemented over any media type, including point-to-point links, multi-access links, and ATM.

### 16.2.2 Tag Switched VPN Network Components

A tag switched network can consist of two types of physical device. The first is the tag edge or provider edge (PE) router, which is situated at the periphery of the tag switched network. The second is the tag or provider (P) switching router that is situated in the core of the tag switched network. The location of the network components can be seen in Fig 16.1.

Typically, customers access the tag switched network via their customer edge (CE) router, which is connected to the PE router using traditional access technologies (e.g. leased line, frame relay or ATM). Multiple customers can share the same PE, as isolation is achieved via the VPN features on the PE. The CE router is a standard router, which does not require tag VPN capabilities.

The PE router examines packets arriving from the CE router and applies a unique tag based on the incoming interface. This tag identifies the packet as belonging to a specific VPN and is used to direct the packet from the source PE to the egress PE. The source PE router forwards the packet to the tag switched core and the P routers. Tagged packets are forwarded through the core based on tags alone. These P routers only have knowledge about the tag core network and not about individual VPNs. For packets leaving the tag switched network, PE routers perform the reverse function, removing tags from packets and then forwarding to the destination CE.

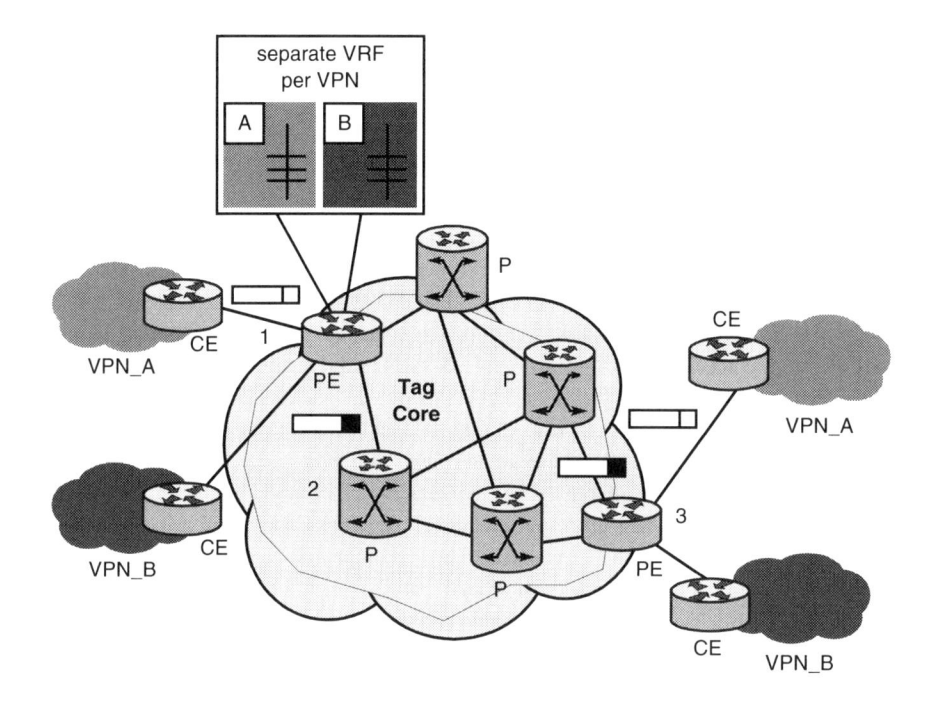

**Fig 16.1** Tag switched VPN.

Since IP datagrams are forwarded within the provider network based on the tag, traffic from different VPNs with conflicting IP address ranges may be safely carried within the same tag switched core.

### 16.2.3 Tag VPN Identification and COINs

The service provider, not the customer, associates one or more VPN community with each customer access interface at initial VPN provisioning. Within the provider network, globally unique route distinguishers (RDs) are associated with every VPN. This ensures that VPN addresses are unique within the routing tables; consequently VPNs cannot be penetrated by 'spoofing' a flow or packet. Users can be part of a VPN only if they reside on the correct physical port and have been assigned the correct RD. Thus tag switched VPNs afford similar levels of security to frame relay, leased line, or ATM services.

It is possible to allow a customer to be a member of multiple VPNs, thus forming an extranet or community of interest network (COIN) between customers. COINs provide a simple and powerful mechanism for the exchange of data between businesses and organisations. For example, car retailers may wish to join a COIN that provides access to on-line parts-ordering systems from manufacturers. COINs allow the specific inclusion of only certain sites into the shared community.

### 16.2.4 End-to-End Tag Switched VPN Routing

Each VPN has its own routing table, known as a virtual route forwarding (VRF) table, residing on the PE. A single PE could contain multiple VRFs; IP addresses only need to be unique within the VRF. The VRF table contains entries for the destination IP addresses reachable within the respective VPN. An extended version of BGP4 carries routing information within a VPN and distributes information about VPNs between PEs. The VRF table maps incoming IP packets from a given site to a pair of tags. One of these tags (the outer) is used in the tag core to forward the packet to the correct egress PE router. The other (inner) tag is used by the egress PE router to forward the packet out of the correct interface and thence to the correct customer. Packets leaving the tag switched network are standard non-tagged datagrams and as such can be forwarded to any standard IP-capable router.

Within the tag switched core, standard routing protocols such as OSPF are operated to establish internal IP forwarding tables. Based on these routing tables, PE and P routers use the tag distribution protocol (TDP) to map the outer tags to these routes. Inner tag mappings are distributed using an extended form of BGP4 for establishing routes between PE routers. TDP does not replace routing protocols but rather uses information learnt from the routing protocol to create tag bindings in the tag switched core.

When a PE router at the entry point of the tag switched network receives a packet for forwarding, the following steps occur. The step numbers are illustrated in Fig 16.1 in section 16.2.2.

1  The PE records on which interface the packet arrived, and therefore which VPN it is in, and then it:

   — analyses the network-layer header including IP destination address,

   — performs appropriate network-layer services such as traffic policing and precedence marking (see later),

   — selects a route for the packet from its VRF table, which was learnt through BGP,

   — applies the inner and outer tags and forwards the packet to the correct PE router.

2  The P router receives the tagged packet and switches the packet to the correct egress interface, based solely on the outer tag, without re-examining the network-layer header. The switching decision is based upon entries in the tag forwarding table, learnt using TDP.

3  The packet reaches the PE router at the egress point of the network. Inner and outer tags are removed and the packet is delivered to the correct destination CE based on the inner tag. A standard, non-encapsulated IPv4 datagram is delivered to the destination CE.

### 16.2.5  Tag VPN Class of Service

Cisco's tag solution provides the ability to map the IP precedence information from incoming IP datagrams into part of the tag header. This information is inserted into the tag header after the inner/outer tags. Packet scheduling and discard policies can be applied within the core based on this information. The value marked in each packet indicates the class of service (CoS) and the policy to be applied. This allows the network to offer customers a number of service classes into which they can elect to place their traffic. Customers can therefore segregate their mission-critical from their non-mission-critical traffic across their VPN.

CoS is provided end-to-end over both access and core networks. A customer can subscribe to either single or multiple classes of service, marking IP frames with the appropriate IP precedence mapping for that level. For instance, one class may provide a guaranteed bandwidth while another class could provide a best effort only.

### 16.2.6  Tag Switching Issues

Issues thet need to be considered include:

- network address translation — this can be implemented by the router at the customer premises (CE);
- resilience within the tag switched core is provided by the dynamic routing protocols which are run within the core, while resilience for the access link can be provided by implementing a secondary (back-up) access with a secondary CE.

## 16.3  IPSec VPNs

### 16.3.1  What is an IPSec VPN?

IPSec is a standards-based framework that provides layer-3 services for confidentiality, privacy, data integrity, authentication and replay prevention.

Although IPSec can be used over a private infrastructure in order to protect data, the use of IPSec tunnels to carry traffic from separate domains over a shared infrastructure creates an IPSec VPN. For instance, IPSec can be used over the Internet, for securing traffic between sites, to allow remote users access to their intranet. It will eventually enable communication between different companies as part of a COIN or extranet.

IPSec is effectively transparent to the network infrastructure and allows an end-to-end security solution via any IP network. Traffic can enter an IPSec VPN via any combination of access technologies, including leased lines, frame relay, ISDN, ATM or simple dial access. Indeed, a key feature of IPSec VPNs is that dial-in access is often provided, allowing roaming users to access network facilities from anywhere in the world without fear of eavesdropping or interference.

IPSec development is carried out within the IETF by the IPSec Working Group. In addition to the security mechanisms defined within the IPSec Working Group documents, additional mechanisms are required to make the system scalable. Specifically the public key infrastructure (PKI) defines how public keys can be signed by certification authorities (CAs) for authentication purposes, and how certificates are managed. Without these features IPSec cannot be widely deployed nor scaled.

### 16.3.2 IPSec Implementation

IPSec functions may be implemented:

- in the IP stack;
- between the IP stack and the network;
- as a separate device or system in a network.

Integration of IPSec into an IP stack (Fig 16.2) requires access to the operating system source code and is applicable to both hosts and security gateways. IPSec will be an inherent function of IPv6; however, there is an increasing number of IPSec-capable IP stacks becoming available, e.g. recent versions of HP's and IBM's Unix.

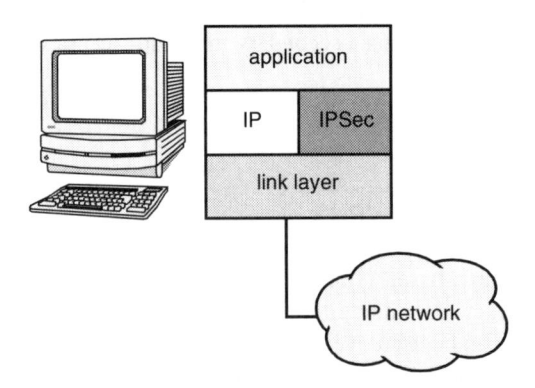

**Fig 16.2**  IPSec integrated into IP stack.

An alternative that does not require access to the IP source code is 'bump-in-the-stack' (BITS) implementations (Fig 16.3). In this case, IPSec is implemented 'underneath' an existing IP protocol stack, i.e. between the native IP layer and the local network drivers. Since the source code for the IP stack is not required, this implementation is appropriate for use with legacy systems. This approach is employed in hosts and gateways because of its simplicity of implementation. However, there are problems with this method, e.g. the increase in packet size added by the IPSec layer can prevent the effective operation of path-MTU discovery.

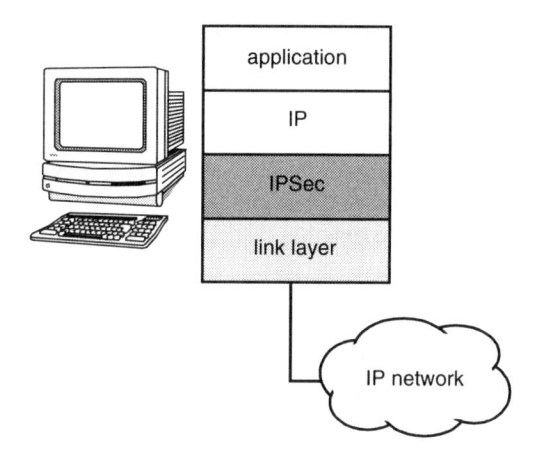

**Fig 16.3**  IPSec 'bump in the stack'.

One particular factor to note with BITS implementations is that it is difficult to write applications which are IPSec-aware, since by its nature BITS implementation renders the IPSec functions invisible.

The use of a physically separate cryptographic processor (Fig 16.4) is a common design feature of network security systems used by the military, and increasingly with commercial systems. It is sometimes referred to as a 'bump-in-the-wire' (BITW) implementation. Such implementations may be designed to serve either as hosts or gateways (or both).

### 16.3.3  IPSec Mechanisms

The security protocols used within IPSec to protect the payloads are:

- authentication header (AH), which provides authentication and protects the payload from modification;

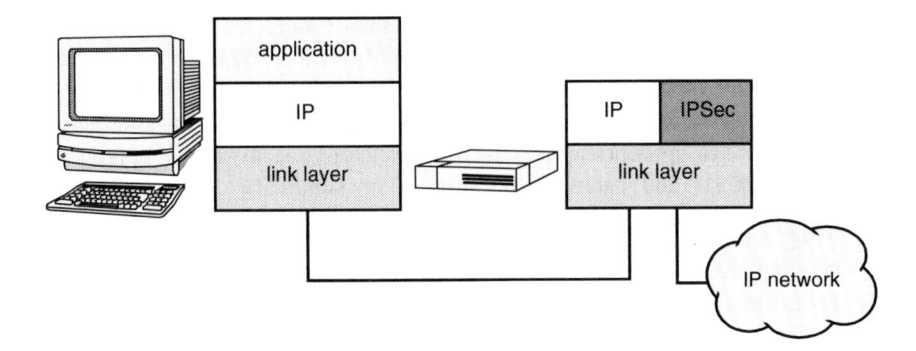

**Fig 16.4** Separate IPSec accelerator.

- encapsulating security payload (ESP), which provides authentication and encrypts the packet, thus preventing both modification and examination of the contents — ESP is the IPSec function that allows VPNs to be implemented.

The security relationships between devices are referred to as security associations (SAs), and the policy is controlled by the security policy database (SPD). The security association database (SAD) maps traffic to the policy, allowing the appropriate level of security to be applied.

Lastly, the key management scheme is Internet key exchange (IKE), although IPSec can be used with other key management schemes.

### 16.3.3.1 Tunnel and Transport Modes

IPSec can operate in two modes — tunnel mode and transport mode (see Figs 16.5 and 16.6). Tunnel mode is used between gateways or between hosts and gateways, while transport mode can only be used between hosts.

In ESP transport mode, only the IP payload is encrypted, and the original IP headers are left intact. This mode has the advantage of adding only a few bytes of overhead to each packet.

Since the original IP header is not encrypted, transport mode allows devices on the public network to see the source and destination of the packet. An attacker would not be able to determine which application sourced the data (i.e. HTTP, SMTP, etc), or to derive any sensitive data from the payload. However, the ability of the network to provide different classes of service, based on the application, is not possible.

Because IP addresses are not hidden, all nodes must have publicly assigned IP addresses. Although ESP in transport mode can be regarded as a VPN (packet contents cannot be examined), it is very inflexible and requires co-operation within the address space.

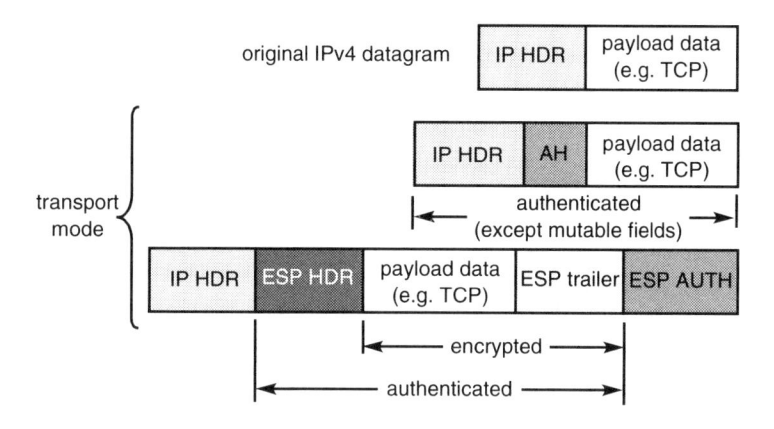

**Fig 16.5** IPSec in transport mode.

In ESP tunnel mode, the entire original IP datagram is encrypted, and it becomes the payload of a new IP packet. This mode allows network devices (e.g. routers and firewalls) to act as IPSec proxies. Such devices provide an encrypted tunnel on behalf of the hosts, as shown in Fig 16.7.

On entry to the tunnel, the source device encrypts packets and forwards them along the IPSec tunnel. The tunnel-end device decrypts the original IP datagram and forwards it to the destination system.

The major advantage of tunnel mode is that the end systems do not need to be modified to enjoy the benefits of IP security.

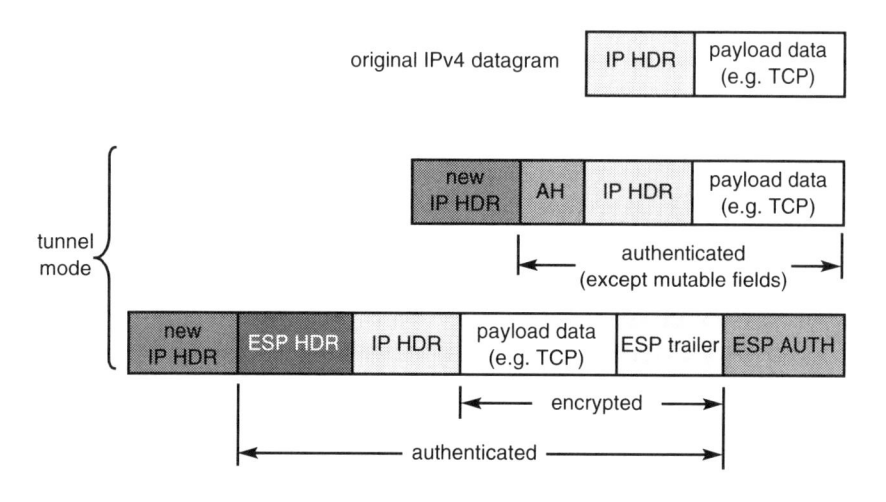

**Fig 16.6** IPSec in tunnel mode.

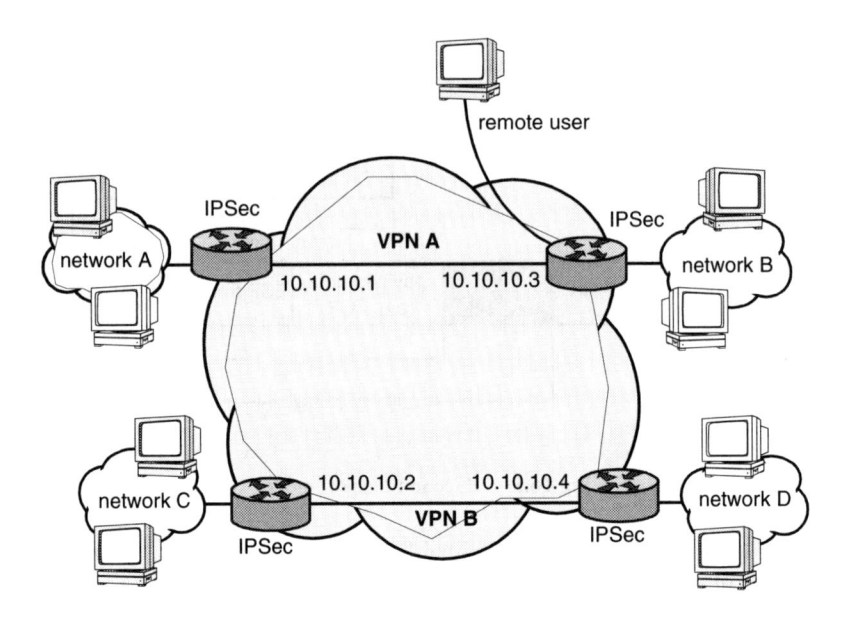

**Fig 16.7** IPSec VPNs.

ESP in tunnel mode also protects against traffic analysis; an attacker can only determine the tunnel end-points and not the true source and destination of the tunnelled packets. Again, class of service functions within the network, that prioritise based on application data, may not be used.

Transport mode may only be used for host-to-host IPSec connections, while tunnel mode may be used for all connection types between hosts and gateways.

### 16.3.3.2 Authentication Header (AH)

The IP authentication header (AH) provides packet integrity and source authentication, and can protect the entire packet's contents against replay.

These three functions are mainly provided using a secret key attached to the original message (for authentication), a hash function (for data integrity) and a sequence number (to guard against replay attacks).

A hash is a one-way function which, when applied to any data, always produces a fixed-length result. If a single bit is modified in the original datagram, at least half of the hash changes, making it simple to detect changes and difficult to create alternative messages with the same hash. In practice, the sender produces a hash of the original message and a secret key (a keyed hash). The receiver then produces their own hash of the datagram and a local copy of the key and compares it to the hash received — the two must be identical for the packet to be authenticated.

It should be noted that AH protects the entire content of an IP datagram except for certain mutable fields of the IP header that are likely to change in transit (e.g. time to live (TTL) and type of service (TOS)).

### 16.3.3.3 Encapsulating Security Payload (ESP)

The IPSec ESP provides data confidentiality (encryption), packet integrity, source authentication, and optional protection against replay. The last three functions are performed in the same way as AH, and are optional.

For encrypting, ESP uses a private key, where both sender and receiver use the same key to encrypt/decrypt the packet. In transport mode, ESP only encrypts the payload. In tunnel mode, a new IP header is added to the datagram, and the original IP header is encrypted, thus hiding the original source and destination addresses.

As ESP cannot encrypt its own IP header (otherwise the packet would become unroutable), it does not therefore protect the IP header. To protect the header, ESP can be used in conjunction with AH, e.g. an ESP datagram nested within an AH datagram.

### 16.3.3.4 Security Associations (SAs)

IPSec provides many options for performing network encryption and authentication. Each IPSec connection can provide either for encryption, for integrity and authenticity, or both. Once the security service (e.g. AH or ESP) is determined, then the two communicating nodes must agree exactly which algorithms to use (e.g. DES or IDEA for encryption, MD5 or SHA for integrity), and keys must be agreed for the session.

The security association is the method that IPSec uses to track all of the particulars concerning a given IPSec session. An SA is a relationship between two or more entities that describes how they use security services to communicate securely.

IPSec SAs require agreement between two entities on a security policy, including:

- encryption algorithm;
- authentication algorithm;
- session keys;
- SA lifetime.

### 16.3.3.5 Security Policy

The security policy dictates factors including:

- what IPSec services are to be used, and in what combination;
- the granularity with which protection is applied (e.g. single device or a range of devices, protocol port numbers);
- the algorithms to use.

### 16.3.3.6 SA and Key Management

Each device in an IPSec network requires some form of key, both to validate the device, and as a basis for initiating encryption. Manual keying is the simplest form of key management to implement but probably the hardest to manage. Given that different keys may be required between every pair of gateways, the difficulty with which keys are managed increases with the square of the number of gateways. More accurately the number of keys, $Kn$, for $n$ gateways is:

$Kn = (n*(n-1))/2$

Thus for 3 gateways, only 3 keys are required. These can be manually entered at each gateway fairly quickly. For 10 gateways this rises to 45 keys and for 100 gateways, to 4950. This illustrates just how unscalable a manual keying solution is, particularly since keys must be regularly updated to maintain security.

The only possible option for a scalable secure network is to use automatic key negotiation. The default for this is the Internet key exchange which is a subset of the ISAKMP/Oakley/SKEME standards.

### 16.3.3.7 IKE SA Authentication Options

Peers within a secure network must be authenticated to each other. IKE is very flexible and supports multiple authentication methods. Entities must agree on a common authentication protocol through a negotiation process. The following mechanisms are described within IPSec.

- Pre-shared keys

  The same key is securely pre-installed on each host. IKE peers authenticate each other by computing and sending a keyed hash of data that includes the pre-shared key. If the receiving peer is able to independently create the same hash using its pre-shared key, it knows that both parties share the same secret, thus authenticating the other party.

- Public key encryption

  Each party has a pair of keys (one private and one public) which are related by a mathematical function. During an IKE negotiation, each party generates a pseudo-random number (a nonce) and encrypts it with the other party's public key. Each party can then use its own private key to recover the nonce. This authenticates the parties to each other.

- Digital signatures

    Public key encryption helps reduce the number of keys and simplifies management. However, there is still an issue with validation of the public keys, i.e. how can a device be sure the public key is sent by the owner of the key. Digital certificates make use of certification authorities as an independent third party to validate the keys and associate them to their owner. This also provides non-repudiation (transactions cannot be denied).

Once authentication of devices is performed, encryption over the IKE tunnel requires a session key to be shared between the tunnel end-points. The Diffie-Hellman key agreement protocol is used to establish this session key.

### 16.3.3.8 IPSec SA Keying Options

When the communicating entities have agreed on which algorithms to use, they must derive keys for IPSec with AH, ESP, or both.

IPSec uses a shared key that is different from IKE. The Diffie-Hellman exchange can be carried out again to derive the IPSec shared key, or it can be based on the existing shared secret generated for the IKE SA (also derived using Diffie-Hellman). The first method provides greater security but is slower.

### 16.3.4  Public Key Infrastructure (PKI)

Public key infrastructures are of interest because they are designed to be both scalable and manageable. They enable the automation of IKE processes.

In order to enrol a device with a CA, it must first generate a public/private key pair. The public key is sent to the CA where, following authorisation, it is signed and transformed into a digital certificate. This certificate may be used by other peers to validate it. The private key must never be disclosed.

In order to manage certificates that are no longer valid, a list of revoked certificates is signed by the CA. This certificate revocation list (CRL) is stored on the CA or in a directory service, and may be accessed by devices when setting up SAs.

### 16.3.5  IPSec Issues

The following points also need to be taken into consideration.

- Interoperability

  Many implementations are currently incompatible, mainly due to imprecise standards. The International Computer Security Agency (ICSA) does provide a measure of approval, but does not guarantee interoperability. Also, the means whereby certificates are enrolled and verified with different certification authorities are unstandardised and there are issues with the propagation of CRLs.

- Network address translation (NAT)

  With transport mode services, NAT may not be performed between IPSec devices using AH or transport-mode ESP because it changes the IP header and authentication fails.

- Resilience

  There are no standardised mechanisms to allow for resilient IPSec connections. Existing hot-standby techniques will not work with IPSec and ways to work around this are required.

- Legal and regulatory issues

  Encryption is still viewed by many countries as munitions, although recently many of these regulations have been relaxed. However, some countries or ISPs will block IPSec traffic as a matter of policy. This could obviously prevent the use of IPSec within a foreign country for remote access.

- Quality of Service (QoS)

  IP QoS mechanisms, using a differentiated services approach, are usually applied according to the marking of the TOS bits in the IP header. Network devices may be configured to apply QoS classification and TOS-remarking policies based on the source and destination IP addresses, and the port numbers (i.e. application information). With ESP, the payload, and sometimes the IP header (in tunnel mode), is encrypted. Hence, the information for QoS classification is no longer available. The way to work around this is to perform packet classification before IPSec processing.

## 16.4 Which Technology?

Typically, legacy private networks have been constructed using ATM, frame relay, SMDS or leased lines. This section positions the emerging IPSec and tag switching VPN technologies relative to legacy technologies, and demonstrates how they can be applied to different scenarios.

### 16.4.1 Legacy VPNs

Legacy intranets are typically built using layer-2 technologies (such as leased lines, frame relay, SMDS and ATM). These technologies are established and are stable, which makes them very attractive. However, there are some disadvantages in comparison with IPSec and/or tag switching.

Layer-2 technology limits flexibility because connections are set up using leased lines or virtual circuits between particular points. Any new site connected to the VPN must have one or more dedicated links to other sites. Because it is based on point-to-point links, the number and size of connections within the overall network needs constant review to ensure efficient routing and bandwidth utilisation. This limits adaptability and can be a barrier to growth. In addition, these connections can be very expensive, especially for long-distance links, thus increasing costs.

Typically, legacy remote users gain access to their intranet using dial-access facilities. This requires the customer to operate arrays of stand-alone equipment for remote access (e.g. modem banks), with teams of skilled management people. IT bills are costly and the system is complex to maintain. Costs are incurred for many telephone lines, and for the long-distance call charges to access the system.

In terms of security, legacy networks provide IP connectivity with no additional security features such as encryption — all traffic is carried in clear text. Also, a firewall is still required when traffic is to be controlled between sites.

With the exception of ATM and to a lesser extent frame relay, most legacy networks do not inherently implement end-to-end packet prioritisation to allow a better usage of the bandwidth (QoS).

Legacy VPNs generally require $N^2$ virtual circuits to be allocated between sites within the network. Each of these VCs will also require configuration depending on customer requirements, e.g. for frame relay the committed information rate (CIR) must be set for every VC. Therefore, fully meshed topologies, for large VPNs, are complex to build and scale.

Lastly, legacy networks have the advantage of being able to carry all layer-3 packets, and hence are not limited to IP.

### 16.4.2 IPSec VPNs

Typically, IPSec VPNs are used over a shared network (e.g. the Internet) for securing traffic between sites, to allow remote users access to their intranet, and will eventually be used for enabling communication between different companies as part of a community of interest network or extranet.

Using IPSec, all sites of a corporate network can communicate securely over the Internet. All that is required is an Internet connection and an IPSec device.

This connection can be set up very quickly and easily, thus providing the flexibility that layer-2 switching technology lacks. If access is only provided via the Internet and not to the Internet itself, no firewall is required, and no intervention by the network service provider is necessary.

Advanced security features can be integrated in an IPSec-based VPN. A common trend of VPN-enabled devices is to integrate firewalling services, as well as other security features such as logging, intrusion detection, URL filtering, virus scanning, etc. So IPSec, combined with other services, can provide a global, secure solution, controlling all traffic down to the users, devices and applications.

Another very attractive feature of IPSec VPNs for organisations is the ability to allow remote users to access their intranet securely.

Indeed, using IPSec-based VPN technology, remote users gain access to the Internet via a local ISP, and then gain connectivity to their intranet via an IPsec VPN tunnel to their gateway. Thus, the enterprises incur relatively cheap local call costs.

Also, with integrated firewall/IPSec VPN solutions, organisations can now have one standard remote access facility, enforcing the overall company security policy at a single point rather then relying on multiple points of access. One has to keep in mind that remote users can usually also gain access to the Internet while on-line, and hence their PC is exposed to Internet hackers. Protection (i.e. desktop encryption or mini-firewall) may be required. The third application of IPSec is extranet support. IPSec, combined with digital certificates and PKIs, provides a secure and scalable solution for the new emerging business model, whereby communication between different organisations is required.

Lastly, IPSec can significantly reduce IT costs. Indeed, for site-to-site VPNs or extranets, IT costs are reduced by replacing existing legacy connections with an Internet connection (or two per site for resilience if required, although there are still issues regarding IPSec and resilience). Also, because it is standards-based, vendor tie-in is minimised — indeed, competition between vendors tends to drive technology forward and equipment prices down. Management costs may also be reduced thanks to the integration of all functionality within a few network components thus allowing centralised management.

IPSec, although not yet fully developed and still facing interoperability issues, is already showing great benefits for eBusiness communications as well as fulfilling the major organisations needs.

### 16.4.3 Tag Switching

Tag switched VPNs provide very similar functionality to legacy private networks, e.g. IP connectivity between sites. However, it provides major advantages, which are greater flexibility and resilience, and in general can be cheaper:

- tag switched VPNs provide a scalable solution — a single link to the core can give the customer a fully meshed network;
- a tag switched VPN can be implemented without any hardware or software alterations or additions to the customer router;
- tag switched VPNs provide support for multiple services such as Internet access, COINs and extranets.

The parameters controlling the configuration of a tag switched VPN are easy to modify, allowing for quick set-up and modification should requirements change.

Tag switched VPNs are built using a shared routed network. They can be built point-to-point (as a legacy VPN) or 'virtually' fully meshed. In the latter, tag switched VPNs make use of the routing protocols of the underlying network, and hence traffic forwarding is more efficient and resilient.

Tag switched VPNs allow for the construction of extranets, provided that all member sites are connected on the tag switched network.

Another major advantage is the inherent support of classes of service. It allows for packet prioritisation for mission-critical applications, and hence provides efficient bandwidth management. Organisations can subscribe to the classes of services they require, and they are guaranteed to get what they pay for.

Tag switched VPNs are relatively cheap compared to legacy private networks. This is essentially due to the use of a shared infrastructure, leading to economies of scale. Also, point-to-point links are no longer required, thus reducing costs for certain network topologies.

One has to keep in mind that, as for legacy networks, tag switched VPNs provide IP connectivity. Security (i.e. encryption, authentication, traffic filtering) can be implemented using additional equipment/features on the organisation's premises or in the network.

### 16.4.4 Legacy Versus New Technology

Table 16.1 summarises the major differences between various VPNs' enabling technologies.

Although IPSec networks are simple to set up and manage for small-medium networks, scaling and management issues currently limit growth in such a way that this technology may not be suitable for large enterprises (e.g. for thousands of connections), be they for separate sites or remote users.

The design and implementation of an IPSec network can greatly affect the overall cost. For remote use, a simple software client is adequate; however, for high-bandwidth point-to-point connections, dedicated hardware will be required, and this must be borne in mind when designing the network.

**Table 16.1**  Comparison of various technologies.

|  | Legacy | IPSec | Tag switched |
|---|---|---|---|
| Flexibility | Low | High flexibility for small-medium networks | Medium |
| Remote users | Stand-alone solution | Integrated | Stand-alone solution |
| Security | Partially, but firewall often required | High (encryption). Can be at user and application level | Partially, but firewall often required |
| QoS | Can be done with certain technologies | Limited by the underlying network QoS | Can be supported end-to-end through the network |
| Protocols | All layer 3 | IP only* | IP only* |
| Cost | Expensive | Competitive | Very competitive |

*Although IPSec and tag switching will only carry IP traffic, there are techniques such as data link switching that allow other protocols to be carried by IP and benefit from the technology.

## 16.5 Conclusions

There are several VPN-enabling technologies, as already mentioned throughout this chapter. However, the overall design of an organisation's networks must take into account the following criteria:

- QoS;
- layer-3 protocols;
- geographical locations;
- performance;
- reliability;
- scalability;
- remote user support;
- flexibility;
- type/level of security;
- cost.

In practice, most organisations will deploy a mixture of several technologies.

Typically, it is foreseen that legacy systems will remain in use for business applications requiring performance and reliability. These networks will slowly migrate to tag switched VPNs to make use of the class of service support, especially if tag switched VPNs prove to be significantly cheaper.

For those organisations that require a flexible global network, together with remote user support and extranets, IPSec is the most suitable solution.

Finally, for tag switched VPNs with remote users or remote sites connected to the Internet, IPSec gateways may be used to provide access to the tag switched core. Also, the two technologies may be combined to provide even more privacy on a tag switched VPN, using IPSec functionality.

## References

1   Rekhter Y et al: 'Cisco Systems' Tag Switching Architecture Overview', IETF RFC 2105 (February 1997).

2   Rosen E and Rekhter Y: 'BGP/MPLS VPN', IETF RFC 2547 (March 1999).

# ACRONYMNS

| | |
|---|---|
| AAA | authentication, authorisation and accounting |
| ABR | area border router |
| ACD | automated call distribution |
| ADM | add-drop multiplexer |
| ADSL | asymmetric digital subscriber line |
| AFRINIC | African Network Information Centre |
| AH | authentication header |
| AIS | accept into service |
| APNIC | Asia Pacific Network Information Centre |
| APON | ATM passive optical network |
| ARIN | American Registry for Internet Numbers |
| ARPANET | Advanced Research Projects Agency Network |
| AS | autonomous system |
| ASDH | access SDH |
| ASN | autonomous system number |
| ASO | Address Supporting Organisation |
| ATM | asynchronous transfer mode |
| ATMP | ascend tunnel management protocol |
| AUP | acceptable use policy |
| BAS | broadband access server |
| BGP | border gateway protocol (iBGP — internal, eBGP — exterior) |
| BITS | bump in the stack |
| BITW | bump in the wire |
| BRAN | broadband radio access network |
| BWA | broadband wireless access |
| B-WLL | broadband wireless local loop |
| CA | certification authority |
| CAR | committed access rate |
| CBR | constraint based routing |
| CBR | constant bit rate |
| CBR-LDP | constraint based routing — label distribution protocol |
| CCL | Cordless Class Licence |
| CDMA | code division multiple access |
| CDV | cell delay variation |

| | |
|---|---|
| CE | customer edge |
| CEF | Cisco express forwarding |
| CIDR | classless interdomain routing |
| CIP | Concert Internet Plus |
| CIR | committed information rate |
| CLEC | competitive local exchange carrier |
| CM | configuration manager |
| CMIP | common management interface protocol |
| COIN | community of interest |
| CORBA | Common Object Request Broker Architecture |
| CoS | class of service |
| CPE | customer premises equipment |
| CPMS | cost per Mbit/s shipped |
| CPU | central processing unit |
| CQC | Concert quality control |
| CRL | certificate revocation list |
| CTMS | Concert trouble management system |
| DAVIC | Digital Audio-Video Council |
| DECT | digital European cordless telephone |
| DES | data encryption standard |
| DHCP | dynamic host configuration protocol |
| DLCI | data link connection identifier |
| DLE | digital local exchange |
| DMS | domain management system |
| DMSU | digital main switching unit |
| DNIS | dialled number information string |
| DNS | domain name server/system |
| DNSO | Domain Name System Supporting Organisation |
| DOCSIS | Data Over Cable Service Interface Specification |
| DPCN | digital private circuit network |
| DS | direct spread |
| DSL | digital subscriber line |
| DSLAM | DSL access multiplexer |
| DTPM | distributed transaction processing monitor |
| DVB | digital video broadcast |
| DWDM | dense wavelength division multiplexing |
| DXC | digital cross-connect |
| EAI | enterprise application integration |
| ECAT | event collection and alarm translation |
| ECMF | equal cost multipath forwarding |
| EDGE | enhanced data rates for GSM evolution |
| EIDS | engineering intelligence database system |
| EJB | Enterprise JavaBeans |

| | |
|---|---|
| EMW | enterprise middleware |
| EPD | early packet discard |
| EPT | equipment planning tool |
| ESP | encapsulating security payload |
| ETSI | European Telecommunications Standards Institute |
| EWF | enterprise workflow |
| FAB | fulfilment, assurance, billing |
| FCC | Federal Communications Commission (USA) |
| FDD | frequency division duplex |
| FDDI | fibre distributed digital interface |
| FEC | forward error correction |
| FH | frequency hopping |
| FITL | fibre in the loop |
| FR | frame relay |
| FSAN | Full Service Access Network (initiative) |
| FTP | file transfer protocol |
| FWA | fixed wireless access |
| GEO | geostationary orbit |
| GGSN | gateway GPRS service node |
| GMSK | Gaussian minimum shift key |
| GoS | grade of service |
| GPRS | general packet radio service |
| GSM | global system for mobile communication |
| GTFM | generic technology fault manager |
| gTLD | generic top level domain |
| GTS | generic traffic shaping |
| HAP | high-altitude platform |
| HDLC | high-level data link control |
| HDTV | high definition TV |
| HFC | hybrid fibre coax |
| HG | home gateway |
| HIPERLAN | high performance local area network (European) |
| HSCSD | high-speed circuit-switched data |
| HTTP | hypertext transfer protocol |
| IA | information architecture |
| IAB | Internet Architecture Board |
| IANA | Internet Assigned Numbers Authority |
| ICANN | Internet Corporation for Assigned Names and Numbers |
| ICP | Internet content provider |
| ICSA | International Computer Security Agency |
| IDEA | Internet design, engineering, and analysis |
| IDSL | ISDN digital subscriber line |
| IEEE802.X | Institute of Electrical & Electronic Engineers LAN standards |

| | |
|---|---|
| IETF | Internet Engineering Task Force |
| IGP | interior gateway protocol |
| IiP | Investors in People |
| IKBS | intelligent knowledge-based system |
| IKE | Internet key exchange |
| IMT | International Mobile Telecommunications |
| INS | integrated network system |
| IOC | Internet operations centre |
| IP | Internet protocol |
| IPC | international private circuit |
| IPSec | IP security |
| IR | information repository |
| IRD | integrated receiver decoder |
| IRTF | Internet Research Task Force |
| ISAKMP | Internet security association and key management protocol |
| ISDN PRI | ISDN primary rate interface |
| ISDN | integrated services digital network |
| IS-IS | intermediate system–intermediate system |
| ISP | Internet service provider |
| ITU | International Telecommunications Union |
| IXP | Internet exchange point |
| JIT | just in time |
| L2F | layer 2 forwarding |
| L2TP | layer 2 tunnelling protocol |
| LAC | L2TP access concentrator |
| LAN | local area network |
| LATNIC | Latin American Network Information Centre |
| LEO | low-earth orbit |
| LES | land earth station |
| LIR | local Internet registry |
| LLU | local loop unbundling |
| LMCS | local multipoint communications system |
| LMDS | local multipoint distribution system |
| LNS | L2TP network server |
| LSA | link state advertisement |
| LSP | label switched path |
| LSR | label switch router |
| LTE | line terminating equipment |
| MAC | media access control |
| MAN | metropolitan area network |
| MC | multicarrier |
| MD5 | message digest 5 |
| MED | multi-exit discriminator |

| | |
|---|---|
| MIB | management information base |
| MMDS | multipoint microwave distribution system |
| MOU | memorandum of understanding |
| MPE | multiprotocol encapsulation |
| MPEG | Moving Picture Experts Group |
| MPLS | multi-protocol label switching |
| MSP | multi-service platform |
| MTBF | mean time between failure |
| MTU | maximum transmission unit |
| NAP | network access point |
| NAS | network access server |
| NAT | network address translation |
| NFM | network fault manager |
| NISM | network inventory and spares management |
| NMC | network management centre |
| NMS | network management system |
| NOC | network operations centre |
| NOU | network operations unit |
| NREN | National Research and Education Network (USA) |
| NSF | National Science Foundation (USA) |
| NTE | network termination equipment |
| OADM | optical add-drop multiplexer |
| OAG | Open Applications Group |
| OAGIS | Open Applications Group Integration Specification |
| OAM | operations and maintenance |
| OCHP | optical channel protection |
| OLO | other licensed operator |
| OLT | optical line termination |
| OMSP | optical multiplex section protection |
| ONU | optical network unit |
| OOB | out of band |
| OSI | open systems interconnection |
| OSPF | open shortest path first |
| OSS | operational support systems |
| PACS | planning, assignment and configuration system |
| PAT | port address translation |
| PCA | percentage calls answered |
| PCMCIA | Personal Computer Memory Card International Association |
| PDA | personal digital assistant |
| PDH | plesiochronous digital hierarchy |
| PE | provider edge |
| PER | power equipment rack |
| PEW | planned engineering works |

| | |
|---|---|
| PKI | public key infrastructure |
| PMP | point-to-multipoint |
| PON | passive optical network |
| PoP | point of presence |
| POS | packet over SONET |
| POTS | plain old telephony system |
| PPD | partial packet discard |
| PPP | point-to-point protocol |
| PPPoA | PPP over ATM |
| PPTP | point-to-point tunnelling protocol |
| PRI | primary rate interface |
| PSO | Protocol Supporting Organisation |
| PSTN | public switched telephone network |
| PTT | public telephone and telegraph (company) |
| PVC | permanent virtual circuit/channel |
| QoS | quality of service |
| QPSK | quadrature phase shift key |
| RA | Radio Agency |
| RADIUS | remote authentication dial-in user service |
| RAG | red, amber and green |
| RD | route distinguisher |
| RFC | Request for Comment |
| RIB | routing information base |
| RIP | routing information protocol |
| RIPE | Réseaux Internet Protocol Européens |
| RIR | regional Internet registry |
| RPC | remote procedure call |
| RSVP | resource reservation protocol |
| RTR | reliable transaction router |
| SA | security association |
| SAD | security association database |
| SB | Systems Business (licence) |
| SDH | synchronous digital hierarchy |
| SDSL | symmetric digital subscriber line |
| SES | Satellite Earth Stations & Systems (ETSI) |
| SFI | special fault investigation |
| SGSN | serving GPRS service node |
| SHA | secure hash algorithm |
| SIM | subscriber identity module |
| SIN | supplier information note |
| SKEME | secure key exchange mechanism |
| SLA | service level agreement |
| SMDS | switched multimegabit data service |

| | |
|---|---|
| SME | small/medium enterprise |
| SMTP | simple mail transfer protocol |
| SNI | service node interface |
| SNMP | simple network management protocol |
| SNOC | secure network operations centre |
| SONET | synchronous optical networking |
| SP | service provider |
| SPD | security policy database |
| SPF | shortest path first |
| SPRing | shared protection ring |
| SS7 | Signalling System No 7 |
| SSB | Supplementary Services Business (licence) |
| SSD | service solution design |
| STM | synchronous transfer mode |
| TC | time and code (division multiple access) |
| TCP | transmission control protocol |
| TDD | time division duplex |
| TDP | tag distribution protocol |
| TE | traffic engineering |
| TLV | type, length, and value |
| TMF | Telemanagement Forum |
| TMN | Telecommunications Management Network |
| TOS | type of service |
| TS | technical services |
| TTL | time to live |
| UAWG | Universal ADSL Working Group |
| UBR | unspecified bit rate |
| UDP | user (unacknowledged) datagram protocol |
| UMTS | Universal Mobile Telecommunications System |
| UNI | user node interface |
| UPS | uninterrupted power supply |
| USB | universal serial bus |
| UTRA | UMTS terrestrial radio access |
| VAS | value added service |
| VC | virtual container |
| VDSL | very high rate digital subscriber line |
| VoD | video on demand |
| VoDSL | voice over DSL |
| VoIP | voice over IP |
| VPN | virtual private network |
| VRF | virtual route forwarding |
| VSAT | very small aperture terminal |
| VToA | voice/telephony over ATM |

| | |
|---|---|
| WAN | wide area network |
| WARC | World Administrative Radio Conference |
| WDM | wavelength division multiplexing |
| WLAN | wireless LAN |
| XC | cross-connection |
| XDMS | cross-domain management system |
| XML | extensible markup language |

# INDEX

Access, *see* Broadband, Code division multiple, Dial, Network, UMTS terrestrial radio, Wireless
Address
    classes (A, B, C, D, E)  218-220
    conservation  223, 229-230
    management tool  227-229
    Supporting Organisation (ASO) 222-223
    translation, *see* Network, Port
African Network Information Centre, *see* AFRINIC
AFRINIC  222
Always-on  107
American Registry for Internet Numbers, *see* ARIN
APNIC  222-224
ARIN  221-223, 225
Ascend tunnel management protocol (ATMP)  146
    *see also* Protocol
Asia Pacific Network Information Centre, *see* APNIC
Asymmetric digital subscriber line  66, 92-94, 97, 99-107, 113-117, 133, 164
ATM passive optical network  114-116
Authentication header  270-273, 275-277
Autonomous system  16, 26, 29, 35, 38, 40, 45-46, 51, 224, 241, 253

Bartholomew S  43
Billington N  91
Bluetooth  119, 132-134, 140
Border gateway protocol  5, 16, 27-41, 45-46, 252-253, 264
    *see also* Protocol
Brain T  55
Broadband  65-73, 91-99, 105-114, 118, 138, 164-166
    access server  110-111

Cable modem  91-98, 118
Capacity management  173, 181-182
Carrier
    scale, *see* Scaling
    tier-1  47, 49, 53
Challinor S  9, 217
Chuter J  143
Classless interdomain routing  5, 219
Code division multiple access  120-122, 134
    *see also* UMTS terrestrial radio access
Collocation  56
Community of interest network  263-266, 277
Configuration management  4, 6, 85, 173, 184, 200
Constraint based routing  237-243, 255
Cordless  119, 129-136, 139-140
Cottage industry, *see* Model
Customer
    inventory database  178
    reception  172, 175-182

## 290  Index

service management   174
Diagnostics
    first-line   172, 176-177, 179-180
    second-line   172, 176-177
Dial access   143-154, 197, 266
Digital
    signature   273
    subscriber line access multiplexer 97, 100-101, 107-108, 111-113
    video broadcast 159
    wrapper 76-77
Digitally Enhanced Cordless Telecommunications (DECT)   120, 129-135
DOCSIS (Data Over Cable Service Interface Specification)   96
Domain name system   18-22, 222-223
Domain Name System Supporting Organisation   223
Dynamic allocation of addresses   230

Encapsulating security payload   268-271
Enhanced data rates for GSM evolution (EDGE)   120-124
Enrico M   91
Enterprise application integration   208-214
Exterior
    gateway protocol   27
        *see also* Border gateway protocol
    routing protocol, *see* Routing protocol
    *see also* Protocol

Fenton C   119
Fibre in the local loop   114-115
Fidler A   157
Fitch M   157
Fixed wireless access, *see* Wireless

G.lite   102-103, 106
Gateway GPRS service node   125
General packet radio service (GPRS)   121-128, 141
Global system for mobile communication (GSM)   120-129, 131, 134, 140

Hanafi N   261
Harris J   119
Hatch C B   197
Hawker I   65
High-speed circuit-switched data   122-123
Hill G   65
Hills S   261
HIPERLAN   130-138
Howcroft A   25

IANA   220-222
IEEE802.11   38, 130-132, 134
Information repository   213-214
Interior
    gateway protocol   27, 34, 235, 237, 248, 260
    routing protocol, *see* Routing protocol *see also* Protocol
International Mobile Telecommunications 2000 (IMT2000)   120
Internet
    Assigned Numbers Authority, *see* IANA
    exchange point   51-54, 56
    key exchange   268, 272-273
    protocol version 4 (IPv4)   107, 113, 230, 265
    protocol version 6 (IPv6)   7, 41, 113, 230, 266
    *see also* IP, Protocol
    security, *see* IPSec
Investors in People   174, 185-188
    *see also* People management

IP
　　address  18-20, 34, 36-37, 109-113, 146, 152-154, 217-230, 264
　　　*see also* Address
　　assignment request  224
　　security, *see* IPSec
　　VPN  106, 261
　　　*see also* VPN
IPSec  265-278
　　mechanism  267-273
　　VPN  265-276
　　　*see also* VPN

Kelly J  91

Latin American Network Information Centre, *see* LATNIC
LATNIC  222
Layer-2 tunnelling protocol (L2TP)  146-147, 152-154
　　*see also* Protocol
Lean production, *see* Model
Leaton R E A  197
Legacy VPN, *see* VPN
Licence exemption  135
Link protection  246-248
LIR  223-230
Local Internet registry, *see* LIR
Local multipoint distribution system  91-92, 96-98, 105-107, 138
McGlaughlin D  261
Meaneaux J  189
Middleware  208-212
Model
　　cottage industry  193-194
　　lean production  193
　　operational  189-195
　　standardised products  193-194
MPLS TE framework, *see* Multiprotocol label switching
Multicast  5-6, 21, 106-107, 161

Multiprotocol label switching 41, 105-107, 235-251
　　TE framework 237-250
　　*see also* Protocol

Network
　　access point  51
　　access server  143-146
　　address translation  112-113, 229-230, 265, 274
　　management 3-5, 83-88, 162-163, 173, 176
　　termination equipment, *see* NTE2000
　　*see also* Optical networking
Nigeon B  119
NTE2000 103-105

Open Applications Group (OAG)  209-211
Operational
　　model, *see* Model
　　process  197-199
　　support systems (OSS) 174-178, 186, 197-214
Operations  172-186, 192-195
Optical networking  68, 74-83
Ozdural J  171

Path selection  238-239, 242-243
Peering policy  29, 54
People management  174
　　*see also* Investors in People
Point-to-point tunnelling protocol (PPTP)  146
　　*see also* Protocol
Port address translation  112, 229
　　*see also* Network
POTS splitter  103-104
PPP over ATM  110-111
Pre-shared key  272
Protection and restoration  81-82

292  *Index*

Protocol
    Supporting Organisation (PSO) 222
        *see also* Ascend tunnel management protocol, Border gateway protocol, Exterior gateway protocol, Exterior routing protocol, Interior gateway protocol, Interior routing protocol, Internet protocol, Layer-2 tunnelling protocol, Multiprotocol label switching, Point-to-point tunnelling protocol, Resource reservation protocol, Routing protocol, Simple network management protocol
Public key encryption   272-273

Q.931   148-149
Quality of service   21, 32, 51, 72-83, 105-107, 113-115, 132, 174-179, 185-186, 250-251, 274-278

Regional Internet registry, *see* RIR
Remote authentication dial-in user service (RADIUS)   144-154
Réseaux IP Européens, see RIPE
Resource reservation protocol (RSVP)   113, 242-246
    *see also* Protocol
Restoration   81-82, 183, 185, 246-247
RIPE   107, 146, 221-22
RIR   220-229
Roberts P A   217
Router   14-15, 25-41, 57, 161-164, 183, 240-260, 262-264
Routing protocol   14-16, 25-29, 241-248

exterior   15-16, 27, 224
interior 15, 28-29
    *see also* Protocol

Satellite   18, 91-94, 98-99, 157-169
Scalable operations challenge   194
Scaling   2, 25-41, 147-151
Security
    association   268-273
    policy database   173, 180, 268-271
    *see also* IPSec
Service
    management   3, 166-167, 203, 205
    surround   171, 185-186
Serving GPRS service node   125
Simple network management protocol   167-168, 176, 180, 207, 213
    *see also* Protocol
Spraggs S   235
Standardised products, *see* Model
Synchronous digital hierarchy   16, 65-88, 173, 260

Tag switched VPN, *see* VPN
Taylor I   65
TD/CDMA, *see* UMTS terrestrial radio access
Technical service (third-line)   172-173, 176-177, 183-184
    *see also* Diagnostics
Telecommunications Management Network (TMN)   166-167, 205-206
Telehousing   56
Telemanagement Forum (TMF)   205
Tier-1 carrier, *see* Carrier
Traffic
    engineering   40-41, 235-259
    trunk attribute   238-240, 245
Tunnel termination   148, 152-154

*see also* Layer-2 tunnelling protocol, Point-to-point tunnelling protocol

UMTS terrestrial radio access (UTRA)
   UTRA TD/CDMA   121-122, 134
   UTRA W-CDMA   121
Universal ADSL Working Group (UAWG)   106
Universal Mobile Telecommunications System (UMTS)   119, 124, 128-134, 166
   *see also* UMTS terrestrial radio access (UTRA)
Utton P C   197

Very high rate digital subscriber line   115-116
Very small aperture terminal   99, 163
Virtual private network, *see* VPN
Voice over DSL (VoDSL)   106
VPN   20, 78, 146, 197, 261-279
   legacy 274-275
   tag switched 261-265, 276-277

Wavelength division multiplexing   17, 65-68, 71, 74-83, 114-115, 235, 260
W-CDMA, *see* UMTS terrestrial radio access
Willis B   119
Willis P J   1
Wireless access   96, 119-141
   fixed access   119, 135-139
   LAN   119, 130-134, 138

Young G   91